THE
DYNAMIC
PLANET

THE
DYNAMIC
PLANET

W. G. ERNST

COLUMBIA UNIVERSITY PRESS

NEW YORK

COLUMBIA UNIVERSITY PRESS
New York   Oxford
Copyright © 1990 Columbia University Press
All rights reserved

Library of Congress Cataloging-in-Publication Data

Ernst, W. G. (Wallace Gary), 1931–
The dynamic planet / W.G. Ernst.
p.   cm.
Includes bibliographical references.
ISBN 0-231-07230-9.   ISBN 0-231-07231-7 (pbk.)
1. Earth.   2. Geodynamics.   I. Title.
QE501.E67   1990
550—dc20   90-30274
                    CIP

Casebound editions of Columbia University Press books are Smyth-sewn
and printed on permanent and durable acid-free paper

Printed in the United States of America
c 10 9 8 7 6 5 4 3 2 1

Test design: SUSAN PHILLIPS

Ⲙ CONTENTS

≡ PREFACE

Twenty years ago, I wrote a brief elementary text on rocks and minerals entitled *Earth Materials;* it was one of a group of short volumes aimed at the advanced college freshman and *Scientific American* audience. Although I was pleased with "Earth Mat," my contribution appeared to have been presented at too high a level to be very effective in an introductory earth science course, yet insufficiently detailed to provide a reasonably complete mineralogy-petrology text for earth science majors. Because it was written prior to the commencement of exploration of the solar system, as well as the full blooming of plate tectonic concepts—both of which have subsequently revolutionized our understanding of the origin and diversity of earth materials—it seems high time to present a next-generation approach. Similar to the earlier effort, familiarity with high school–level chemistry is assumed. Just in case it is needed, the periodic table of the elements is reprinted at the back of the book.

Although patterned after the original in the mineral and rock sections, the present book represents a considerable departure from the treatment in "Earth Mat." The title, *The Dynamic Planet,* alludes to the inexorable constructional forces internal to the planet that couple with degradational processes at the surface. Together, they produce a unique life-giving habitat within the inner solar system. Sections dealing with the Earth's place in the solar system, and sections dealing with its internal structure, have been added and expanded, respectively; new chapters deal with Alfred Wegener, continental drift and plate tectonics, with energy and mineral resources, with geologic hazards and the Circumpacific "Ring of Fire," and with the origin and evolution of the Earth's crust. Many terms are defined in the text, where they constitute the subject under.consideration. Other, more tangential expressions are identified parenthetically. A glossary has been added after chapter 7 for further reference if the terminology becomes confusing.

Considerably reduced and revised are the chapters describing minerals and rocks, and mercifully eliminated is the chapter that introduced classical thermodynamics and phase equilibria. The latter subject rightfully belongs in a modern mineralogy-petrology text, but this is not what I have written. Nor have I put together an all-encompassing, modern earth science text; several excellent elementary books already exist. Instead, I have attempted to show how the exciting developments in space exploration and, especially, global tectonics are impacting the solid earth sciences. Far from a comprehensive, beginning geology text, *The Dynamic Planet,"* pitched at a higher level as it is, contains what I regard as some of the central, provocative, and transformative, themes in this ever-expanding discipline.

⎓ ACKNOWLEDGMENTS

I would like to thank several people at UCLA: Harriet Arnoff, who typed and retyped the manuscript, Jean Sells, who drafted many of the illustrations, and my faculty associates Mark D. Barton and Wayne A. Dollase, who helpfully reviewed portions of it. Half a dozen different elementary classes at UCLA were subjected to the presentation which was evolved into this book: for their anguished but constructive feedback I am also very grateful.

THE
DYNAMIC
PLANET

≡1
≡STRUCTURE OF THE EARTH AND ITS
≡DYNAMIC CHARACTER

As a function of their orbital distances from the Sun, the planets in our solar system exhibit systematic and progressive variations in their chemical constitution and surface appearance. Spatial relationships of the nine planets, as well as those of a typical comet (a wandering "dirty snowball") and of the asteroid belt (meteorite swarm), are shown in figure 1.1. Table 1.1 presents additional solar system information in more quantitative fashion. The innermost planet, Mercury, is metal-rich; the next three, Venus, Earth, and Mars, fundamentally stony, with the outer planets representing progressively more gassy giants and, finally, proportionately more icy, condensed bodies. The Oort cloud, at the outer gravitational limit of our solar system, is the ill-defined home of icy condensates, or comets.

The reason for these contrasts in planetary composition is not known with absolute certainty, but appears to be related mainly to distance from the Sun. Condensation of the solar nebula 4.5–4.6 billion years ago (see chapter 7) might have led to initial differences in the composition of the planets during their growth stages, reflecting chemical heterogeneities in the primordial dust cloud. On the other hand, the solar wind (energy radiating from the Sun) undoubtedly has exerted a strong influence over the distribution of the more volatile elements in the planets; those bodies closer to the Sun each contain high proportions of normally solid (refractory) elements of low volatility whereas, proceeding away from the Sun, the condensed masses become progressively enriched in low-condensation and low-melting temperature (gassy) elements. Very probably, this thermal gradient (temperature gradient), encompassing

1

the Sun and its environment has been largely responsible for the compositional contrasts of the planets.

In fact, each planet is unique in its external aspect and internal nature. The Earth is the middle member of the three stony planets. These terrestrial-like bodies, and our earthly satellite, the Moon, exhibit different surficial features principally because of the contrasting degrees to which their outermost rinds have been reworked. This subject will be taken up again in chapter 7, but suffice it to note here that the interiors of all the planets are hotter than their surfaces because of internal (primordial, radioactive, self-compressive, and kinetically generated) heat. The most efficient mechanism for removal of large quantities of buried heat is by material circulation, or overturn, within the planet, with cooling (energy loss) taking place at the surface. Larger masses contain more heat, and—possessing larger mass/surface ratios—rid themselves of thermal energy more slowly; hence they remain hot and internally turbulent for a longer period of time.

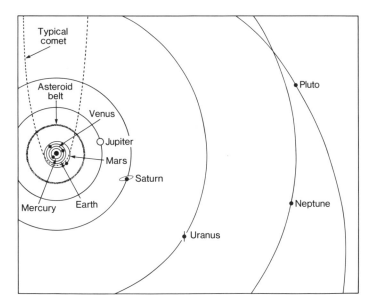

FIGURE 1.1. Orbits of the nine planets in the solar system, as well as that of a tyical comet in Sun-approaching orbit, and of the asteroid belt, properly scaled (after Hartman 1983, figure 2.1). Some planetary satellites, such as the Earth's Moon, the four Galilean satellites of Jupiter, Saturn's Titan, Neptune's Triton, and the largest asteroids are bigger than, or approach the sizes of Pluto and Mercury, the smallest planets. Pluto has an elliptical orbit, and, while generally farther from the Sun, crosses the orbit of the eighth planet, Neptune.

TABLE 1.1. Distances from the Sun, Diameters, and Types of Planets in the Solar System

	Distance from Sun in AU [a]	Diameter of Planet in km	Planetary Type
Mercury	0.4	4,878	iron
Venus	0.7	12,104	stony
Earth	1.0	12,756	stony
Moon	1.0	3,476	stony
Mars	1.5	6,796	stony
Asteroids	~2.8	small	stony
Jupiter	5.2	142,796	gassy
Saturn	9.5	120,660	gassy
Uranus	19.2	50,800	gassy/icy
Neptune	30.0	48,600	icy
Pluto	39.4	2,400–3,800	icy
(Oort Cloud)	100–50,000	small	icy

[a] One astronomical unit (AU) is the distance between the Earth and Sun, about 149,600,000 km.

Assuming that meteorite populations are (and were) relatively uniformly distributed throughout our portion of the solar system, crater density on a planet provides a relative measure of the extent of internally fueled surface reworking: the more numerous the impact craters, the slower the bodily circulation and consequent surficial reworking by mass flow within the planet. Intensely pockmarked planets are nearly dead, or at least are less active internally than those lacking an abundance of such topographic phenomena. Thus, if we compare the relative sizes (and consequent heat contents) of the Moon, Mercury, Mars, Venus, and the Earth with their external features, it is apparent that most of our planet's surface has formed during the very recent geologic past, whereas the Moon, Mercury, and Mars became almost totally inactive several billion years ago. Venus is far less well known because of its dense, obscuring CO_2 cloud cover, but radar imagery has provided enough topographic control to conclude tentatively that surface activity—hence internal circulation—probably ceased there more than a billion years ago. These relationships are illustrated diagrammatically in figure 1.2.

Thus, although possessing chemical characteristics similar to its sunward, hotter sister Venus, as well as its outer, colder brother Mars, the Earth is markedly different from these neighboring siblings. This is understandable given the relative sizes and internal heat budgets discussed above. But it is also a consequence of the position of our planet relative to the radiant solar flux (solar wind). Solid, liquid and gaseous H_2O are stable at and near the Earth's surface. Venus' proximity to the

Sun is responsible for its higher surface temperatures and lack of liquid H_2O oceans. Mars, in contrast, is so distant from the Sun that, on its surface, H_2O frost is presently confined to seasonal polar caps (which actually consist mostly of dry ice—solid CO_2). We need only to glance at photos taken from spacecrafts (e.g., compare figures 1.3, 1.4 and 1.5) to recognize that, in comparison to the dense, torrid, carbon dioxide–enshrouded Venus, and the arid, frigid, nearly atmosphereless Mars, the Earth's outer envelope consists of a particularly hospitable and dynamic

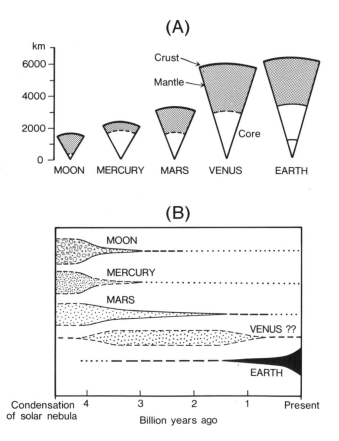

FIGURE 1.2. Relative (A) sizes and gross structures of the inner, terrestrial planets, and of the Moon, and (B) inferred surface reworking of these bodies due to volcanism, erosion, deposition, and meteorite bombardment (modified after Head and Solomon 1981, figures 3 and 12). The increasing surety of time of surface feature production is shown by dotted, dashed, and solid lines, respectively. The vertical axis in (B) is a spatial separation parameter that has no particular significance.

interplay among atmosphere (gassy sphere), hydrosphere (watery sphere), and variegated surface of the solid planet. We have much to love about planet Earth.

Firmly on the ground, we may suspect that, contrary to popular belief, mountains and rivers of our terrestrial landscape are not the immutable, permanent features they seem to be. The forces of erosion (wind, running water, glaciation, and down-slope mass movement), acting over the millennia are reducing topographic prominences to more subdued outlines. Rivers, flowing in their ever-shifting courses, carry solid debris and other, dissolved materials supplied by the agents of erosion toward an ultimate rendezvous with the sea. There, the entrained particles spread laterally and settle, and dissolve constituents eventually precipitate out. Over the billions of years of geologic time since the formation of the Earth (see chapter 7), these processes must have planed off the continents many times over, and deposited vast quantities of sediment in the ocean basins.

Why, then, isn't the Earth a nearly featureless plain at or beneath sea level? Experience tells us that this is not at all the case; graphic proof is demonstrated by a map of the Earth's surface, such as illustrated in figure 1.6. We will pose this question again, and find a fuller explanation in chapter 2. Apparently, constructional forces within the Earth, driven by the slow but inexorable motions of deeply buried solid

FIGURE 1.3. Mariner 10 photo of Venus from a distance of 720,000 kilometers (NASA photo).

FIGURE 1.4. Apollo Spacecraft photo of the (A) Earth and (B) Moon (plus Earth) at different scales (NASA photos).

(A)

(B)

but plastic rock, have swept the ocean basins nearly free of sediments and have resulted in upheaval of rocky sections of the Earth's crust. These internal terrestrial processes thus oppose the surface degradation visible to us all, for clearly we still have mountain belts, just as evidently existed in the ancient geologic past.

To develop an appreciation for the processes at work today, and by inference, those that operated during the distant past, we turn to a

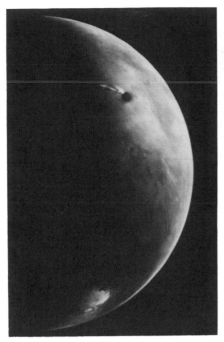

(B)

(A)

FIGURE 1.5. Viking Orbiter 2 photos of Mars (NASA photos). (A) At the top, with CO_2-ice on its western flank is the giant Martian volcano, Ascreans Mons. The large, frost-encrusted crater basin, Argyre, is located near the south pole. (B) General view of cratered surface, original showing the red color of Mars.

consideration of the internal structure of the Earth. Our planet's constitution is a reflection of its original accretionary growth history and subsequent evolution. Such changes over time may largely be attributed to the escape of heat from the deep interior, as will be discussed in chapter 7. But what is the present, generalized structure of the planet? The Earth consists of a series of nested shells, or layers, of contrasting physical and chemical properties. First, the overall structural relationships will be described; then the kinds of evidence employed by earth scientists to deduce the nature of these shells will be presented.

The metric system will be used throughout this volume. As an aid for those steeped in the archaic English measurement system, a few of the more common conversions are presented in table 1.2. To help with the chemistry, the periodic table of the elements is printed at the back of the book.

GROSS INTERNAL CONSTITUTION OF THE EARTH

Diagrammatic sketches, representing a sector from the atmosphere to the center of the Earth are shown in figure 1.7. The Earth is sur-

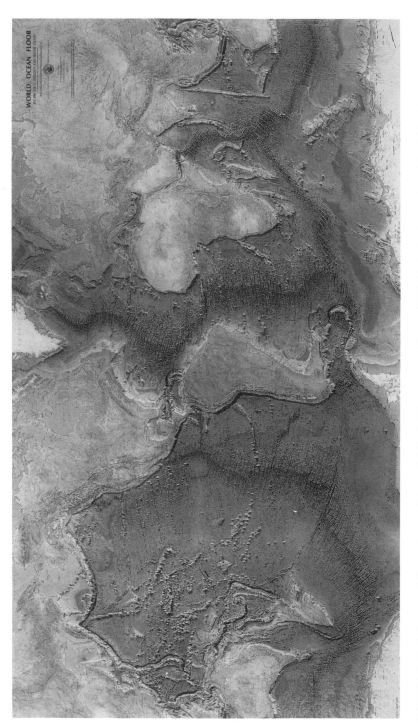

FIGURE 1.6. Relief map of the surface of the solid Earth (Heezen and Tharp 1977). Note that our planet is divided into two major physiographic provinces—continents and ocean basins. The latter are full "to the brim" with seawater, and in fact overly so, for the continental shelves are portions of the high-standing continents and islands that are slightly below sea level.

TABLE 1.2. Selected Conversion Factors

Metric		*English*	
1 bar	= 0.9869 atmosphere	1 atmosphere	= 14.696 pounds/square
	= 10^5 pascals		inch
	= 100 kilopascals		= 1.0133 bars
1 kilobar	= 1,000 bars		
	= 100 megapascals		
1 megabar	= 10^6 bars		
	= 100 gigapascals		
Temperature in degrees Celsius (T_C)		Temperature in degrees Fahrenheit (T_F)	
	= $5/9(T_F - 32)$		= $9/5 T_C + 32$
Temperature in degrees Kelvin (T_K)			
	= $T_C - 273.16$		
1 centimeter	= 0.3937 inch	1 inch	= 2.540 centimeters
			= 25.40 millimeters
1 angstrom (Å) = 10^{-8} centimeter		1 foot	= 0.3048 meter

rounded by a gaseous layer consisting chiefly of nitrogen and oxygen (N_2 makes up about 79 percent, O_2 about 20 percent of the atmosphere, with CO_2 and other, rarer gas species constituting the remainder); the atmosphere is densest at sea level, and becomes progressively rarified with elevation. No sharp break with outer space exists, but the gas molecules are so widely dispersed above about twenty kilometers that we may regard this elevation as the effective limit of the earth-enveloping atmosphere. The ocean basins make up about two-thirds of the surface of the solid Earth, and are great depressions lying about five kilometers below sea level. The continents and islands, together constituting about one-third of the solid Earth's surface, rise above sea level only a few hundred meters on the average. Because seawater more than fills the ocean basins, the hydrosphere laps onto the margins of the continents and islands (see, for instance Hudson's Bay; the Grand Banks, and the East China Sea); these are the so-called continental shelves. Thus, the Earth's surface consists mainly of ocean water, making up slightly more than 70 percent of the exposed area; land, some of which is lake- or ice-covered, constitutes the remaining 30 percent.

The surficial distinction between continents and ocean basins seen in figure 1.6 reflects an underlying structural and chemical difference clarified in figure 1.7B. The outer rind of the solid Earth, called the crust, is of two distinctly different types, continental and oceanic. Continental crust, enriched in silica, alkalies, volatiles, and radioactive elements, is characterized by light-colored granitic and allied rocks (see chapter 4

(A)

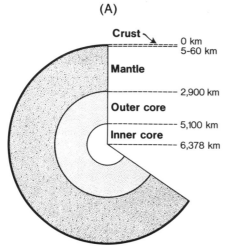

FIGURE 1.7. Schematic section through the Earth. The overall structure, properly scaled, is illustrated in (A). Details of the crust and mantle are shown schematically (not to scale!) in (B). The Mohorovicic discontinuity is abbreviated M, the narrow zone at the core-mantle discontinuity as D″.

(B)

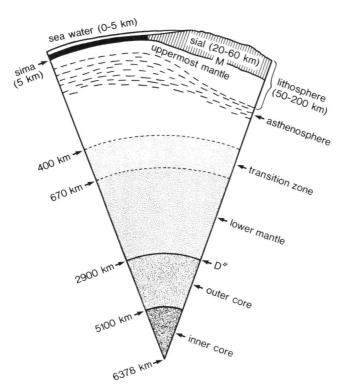

TABLE 1.3. Comparison of the Estimated Compositions in Weight Percentages of the Average Upper Continental Crust, Oceanic Crust, and Model Upper Mantle

Oxide	Average Mantle[a]	Average Sima[b]	Average Sial[c]
SiO_2	45.1	49.6	66.4
TiO_2	0.2	1.4	0.6
Al_2O_3	4.6	16.1	15.5
Fe_2O_3	0.3	—	1.8
FeO	7.6	10.2	2.8
MgO	38.1	7.7	2.0
CaO	3.1	11.3	3.8
Na_2O	0.4	2.8	3.5
K_2O	0.02	0.2	3.3
P_2O_5	0.02	—	0.2

[a] Ringwood (1975).
[b] Pearce (1976); total iron cast as FeO.
[c] Poldervaart (1955).

for a description of these rock types), and is about twenty to sixty kilometers in thickness. It is called sial because it is rich in silicon and aluminum. Oceanic crust, enriched in calcium, iron, and magnesium, contains only modest amounts of potassium and sodium, and lesser amounts of silicon when compared with continental crust. It is almost exclusively basaltic in composition (see chapter 4), and is approximately five kilometers thick. We term the oceanic crust sima, reflecting its major constituents, silicon and magnesium. In aggregate, the crust is a very thin skin, representing a little less than 1 percent of the whole Earth by volume, and only about 0.4 percent by mass (weight). Average compositions of the crust and the underlying mantle material are presented in table 1.3 (compare also with table 4.2).

The boundary between the crust and the mantle, located beneath both the continents and ocean basins, is known as the Mohorovicic discontinuity—Moho or M, for short. It is approximately planar, and in most cases nearly horizontal. This discontinuity probably is a zone of finite thickness, but the width is no more than a few hundred meters under the oceanic crust, perhaps somewhat greater under the continents. The physical and chemical properties of the underlying mantle contrast markedly with those of the crust above.

The mantle in its entirety represents the largest proportion of the Earth, some 83 percent by volume, and 68 percent by mass. Relative to the crust, it is rich in magnesium, less so in iron and silicon, and is depleted in volatile and radioactive elements. The mantle consists of a number of distinct layers. The uppermost is relatively rigid, and, along with the overlying, comparably rigid crust, is termed the lithosphere (rocky sphere). Beneath old portions of the continents, the lithosphere may be several hundred kilometers thick, but within ocean basins it may be as thin as fifty kilometers, and locally is even thinner. The lithosphere, both mantle and crust, is divided into seven gigantic plates and numerous smaller ones (not illustrated in figure 1.7), each of which moves as a coherent, integral unit relative to the other plates. The individual plates are shown in figure 2.13. Beneath the lithosphere lies a soft, relatively weak mantle layer know as the asthenosphere (glassy or weak sphere). Its upper boundary with the lithosphere is relatively sharp on a scale of several kilometers, but it passes downward by degrees into stiffer, more rigid mantle at depths on the order of 220 kilometers or more. The so-called upper mantle consists of lithosphere plus asthenosphere plus more rigid mantle material down to depths of 300–400 kilometers. From below about 400 to approximately 670 kilometers is situated the transition zone of the upper mantle. In this region, physical properties of the upper mantle seem to change fairly rapidly and continuously with increasing depth. The gradual emergence of nearly constant physical characteristics at depths exceeding those of the transition zone signals passage into the so-called lower mantle. The latter appears to be relatively homogeneous down to a depth of about 2,900 kilometers at the boundary with the Earth's core.

The core-mantle boundary, and a lowermost mantle zone abbreviated D″ by geophysicists, represents a zone of profound change in the constitution of the materials that make up the Earth. The overlying mantle shells consist of relatively low-density magnesium and silicon oxide phases, whereas the outer core at depths between 2,900 and 5,100 kilometers consists of liquid iron-nickel alloy (mostly iron). From a depth of 5,100 kilometers to the Earth's center at about 6,378 kilometers, the inner core consists of a solid iron-nickel alloy (again, mostly iron). The outer, liquid core makes sup about 94 percent of the core or 15 percent of the Earth by volume, with the inner, solid core constituting the remaining 6 percent of the core or 1 volume percent of the whole Earth. Because of its great density, the core represents more than 31 percent of the Earth by mass.

⚏ EVIDENCE BEARING ON THE STRUCTURE OF THE EARTH

The overall layering of our planet, as just presented, has been deduced employing numerous lines of evidence. Man has the ability to sink mine shafts (three kilometers) and drill holes (thirteen kilometers) but a short distance into the continental crust, less deeply into the ocean basins. Accordingly, most of the data used to decipher earth structures are model dependent and circumstantial, and involve remote sensing techniques. How can we be sure that the inferred architecture is correct? The evidence supporting the structure of the Earth as sketched above is, in fact, quite impressive because of its diversity and consistency.

Seismic Data

Much of what is known regarding the internal constitution of the Earth arises from study of the nature of propagation of seismic waves. The transmission velocity of vibrational energy, such as results from an earthquake or a man-made explosion, is a function of the chemical and mineralogic composition of the medium as well as its temperature, pressure, and physical state. In general, transmission velocity increases with decreasing mean atomic number (see chapter 3) and decreasing temperature, and with elevated pressure. The overall consequence is that, at greater depths in the Earth (hence higher pressures), the speed of propagation of seismic energy within any layer of constant chemical/mineralogical constitution increases. Thus the velocity of seismic waves increases downward. The transmission velocity through a substance decreases, however, when that substance melts.

Several different types of seismic wave are generated (as well as heat) by a release of energy within the Earth. Part of the vibrational energy is transmitted along the surface as relatively slow surface waves, while the rest of the shock-induced mechanical energy passes through the Earth as body-transecting waves. Two types of the latter may be distinguished: compressional (P) waves and shear (S) waves (see figure 1.8). P waves involve the relatively rapid propagation of successive compressions and rarefactions through a medium, whereas S waves represent vibrations transverse to the direction of travel, at intermediate velocities.

Seismographs are instruments, located at specific observation stations, that record the nature of vibrations transmitted through the Earth and their intensities. By using the information supplied by two or more seismograph stations, the average rate of passage of energy through the intervening geologic medium can be obtained. Where a discontinuity, or change in physical and/or chemical parameters (and therefore transmission properties of adjacent layers), is encountered, part of the seismic energy is reflected (bounced off), whereas the rest is refracted (bent) into the newly penetrated layer. Reflection of vibrational energy is good evidence that a discontinuity exists at depth, whereas the speeds of transmission of the several types of wave provide constraints regarding the chemical and mineralogic constitution of the various layered media.

Consider, for instance, the nearly horizontal boundary between the Earth's crust and mantle: the Mohorovicic discontinuity. The reflection and refraction of seismic waves provides information concerning the nature of this discontinuity. Much vibrational energy is reflected at this boundary, demonstrating the relatively sharp density contrast across it. The difference in P-wave transmission velocities of the material beneath the Moho (mantle V_p equals 8.0 kilometers per second) versus that in the overlying crust (deep continental V_p equals 6.5 kilometers per second; deep oceanic V_p equals 7.0 kilometers per second) is compatible with the idea that the oceanic crust is largely basalt, the continental crust represents a heterogeneous mixture of gabbro, granite, and amphibolite, and the uppermost mantle is periodotite. (For a description of these rock types, refer to chapter 4.)

Let us now examine the variation with depth of compressional and shear wave velocities, for these data provide the most important constraints on the nature of deep continental crust, the mantle, and the core. Observations around the world at seismograph stations have produced the overall relationships presented in figure 1.9. The crust is too thin to be shown in this diagram. Although a general increase in speed of propagation with depth is evident, there are clear discontinuities and changes in the rate of increase, which must be the result of contrasts in mineralogy, bulk composition, or state of aggregation. At the base of the lithosphere, a slight decrease in V_p and a distinct decrease in V_s is

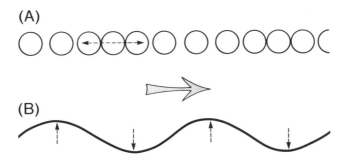

FIGURE 1.8. Distinction between primary (P) compressional waves (under A) and secondary (S) shear waves (under B), two forms of virbrational energy. Transmission direction is to the right (stippled arrow). Note that a unit volume compresses and expands as the P wave passes through, whereas the unit volume shears as the S wave is propagated through it. Both P and S are body waves, in contrast to surface waves; the latter are propagated only along the surface of the Earth. P and S are transmitted through solids, but S waves are extinguished in a liquid medium.

evident. Not illustrated on this figure is the fact that in the mantle region of low velocity, the attenuation (or decrease in amplitude during transmission) of seismic energy also increases markedly, especially for shear waves. These observations are compatible with the hypothesis that the Earth's geotherm—the covariation of temperature and pressure as a function of depth within the planet—has intersected physical conditions appropriate for the onset of incipient melting (the bare beginning of fusion). For this reason, it is supposed that the grain boundaries of mantle materials (i.e., the surfaces of the individual, interlocking crystals) are wetted by a thin film of molten silicate. Such a model would also account for the observed weakness of the underlying asthenosphere relative to the overlying, cooler, more rigid lithospheric plates; the presence of the liquid along grain boundaries would tend to lubricate the hotter asthenospheric material and increase its ductility. Below a depth of about 220 kilometers the upper mantle tends by degree to behave more rigidly, and the rate of attenuation of seismic energy gradually lessens.

Continuing downward, the region between about 400 and 670 kilometers is characterized by mantle materials in which the propagation rate for seismic waves markedly increases with depth. This transition

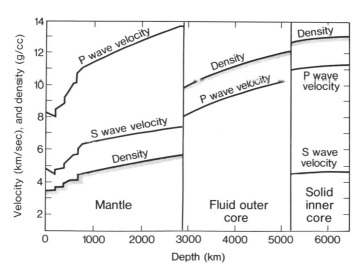

FIGURE 1.9. Densities of Earth shells and transmission velocities of body waves through them as a function of depth within the Earth (National Research Council). As evident from this figure, S waves do not pass through the outer core of the Earth.

zone of the upper mantle is the site where magnesium (and iron) silicates are thought to transform to considerably denser oxide phases that have intrinsically higher V_p and V_s transmission values than the overlying, less dense mineral assemblages. At depths greater than about 670 kilometers, we pass into a more homogeneous lower mantle distinguished by seismic velocities that increase smoothly, gradually, almost imperceptibly. Such changes suggest simple compression of the high-pressure minerals encountered in descending through the overlying transition zone. Many earth scientists consider that, in accord with solar-system abundances of the elements, the lower mantle is richer in its iron/magnesium ratio compared with the upper mantle.

A profound discontinuity exists at the outer core–mantle boundary at approximately 2,900 kilometers depth: V_p drops from about fourteen to about eight kilometers per second, and the shear waves are dramatically extinguished. The latter observation indicates that the outer core must be molten, for S waves cannot pass through a liquid. The decrease in compressional wave velocity is compatible with a relatively high mean atomic number; in addition, the measured speed-of-sound (compressional) waves through liquid iron (plus minor quantities of nickel) at high pressures fit the observed V_p values quite well. An upward jump in compressional wave velocities marks the outer-inner core boundary at about 5,100 kilometers depth. Moreover, within the inner core, shear waves have been detected. These two observations taken together indicate that the inner core of the Earth is solid, probably dominantly iron (plus minor nickel).

Not shown in figures 1.7 and 1.9 are recent discoveries that the mantle is not only vertically layered, but is also laterally heterogeneous. That is, different propagation velocities of seismic waves in various regions indicate chemical/mineralogic and/or thermal contrast. Variation of wave velocity in different directions at a single depth location indicates anisotropy of the material in the mantle as well. These differences reflect some, as yet unknown, combination of chemical and temperature contrasts, but their existence implies differential flow at various levels within the deep Earth.

Understanding the seismic evidence is critical for an appreciation of the layered nature of the Earth. Clearly, our planet consists of a series of nested shells, but can we be sure that other materials would not equally well satisfy the observed V_p and V_s behavior? To reduce and/or eliminate such ambiguities, let us turn to a consideration of other phenomena. In aggregate, these additional observations provide exceptionally strong support and allow refinement of the concentric shell Earth model already introduced.

Specific Gravity Data

As determined from measurement of the gravitational constant and the Earth's surface gravitational acceleration and volume, it is known that the specific gravity of our planet is 5.5. (Specific gravity is defined as the mass of an object divided by an equivalent volume of water measured at atmospheric pressure and four degrees Celsius; it is a dimensionless number. Thus water itself has a specific gravity of 1.0. A related term, density, is defined as mass per unit volume; in the metric system, its dimensions are grams per cubic centimeter.)

In addition, the Earth's moment of inertia (rotational inertia), a measure of mass distribution times the square of the effective radius, is less than that of a homogeneous sphere, suggesting that mass is concentrated toward the center of the planet. This relationship can be understood by considering the contrasting specific gravities of the various layers of the planet. The volumetrically insignificant crust has a measured specific gravity of 2.6–3.0, and the major volume of the Earth—the mantle—possesses a specific gravity ranging from 3.2 measured on materials derived from just below the Mohorovicic discontinuity, grading downward monotonically toward an estimated value of 4.6 for a mixture of silicon oxide, magnesium oxide, and iron oxide at the mantle-core boundary; thus the core itself must have an average specific gravity of approximatelly 10–11 in order for the entire Earth to average 5.5. Extrapolated to core temperatures and pressures, such specific gravities are appropriate for iron-plus-minor-nickel alloys, including small amounts of a dissolved low-density solute (silicon, sulfur, or, less probably, potassium). Thus, specific gravity and inertial considerations corroborate the proposed layering of the Earth, with progressively denser units residing in the successively deeper portions of the planet.

Meteoritic Data

Meteorites represent the most complete sampling of extraterrestrial planetary materials from our solar system. Judging by measured orbital trajectories, the vast majority appear to have come from the asteroid belt, a swarm of planetesimals (meteorites and small planetary bodies) circling the Sun in orbits between Mars and Jupiter. Meteorites are of two principal types, stony and iron. The latter are made up predominantly of iron-rich iron-nickel alloy, but minor magnesium silicates occur in some. The former consist chiefly of the magnesium-rich silicates olivine and orthopyroxene (see chapter 3 for brief descriptions of the silicon-oxygen polymerization and structures of these minerals), but many contain minor amounts of iron-nickel alloy and/or sulfides. Most stony meteorites contain numerous spheroidal silicate blebs (small, round particles) termed chondrules. These droplets seem to have coalesced from the solar nebula (the primordial gas cloud) to form accreting planetesimal bodies, appropriately known as chondrites. Stony meteorites of a very rare class, the carbonaceous chondrites, contain hydrous minerals and carbonaceous matter. Does the latter represent the former existence of life in a crustal environment of a now-disrupted planet? Probably not, judging from abiotic isotopic abundances (see chapter 3), but no one knows for sure. The association of organic-like material with H_2O-bearing substances is certainly reminiscent of the terrestrial surface.

The different groups of meteorites are chemically and physically similar to the various shells or layers that have been postulated to make up the Earth. The exceedingly rare carbonaceous chondrites are analogous to the volumetrically insignificant crust. The mantle is thought to be quite comparable to stony meteorites, and the core to iron meteorites. But what about their relative proportions? Two thirds of all meteorites found on the surface of the Earth (defined as "finds") are irons, only one third are stony; but then, whereas iron meteorites are unmistakable, most stony meteorites display the appearance of rather common (even unattractive) terrestrial rocks, hence probably are not recognized as meteorites when accidentally encountered by lay people and off-duty scientists as well. If we consider only meteorites that are retrieved after having been observed to fall (defined as "falls"), we discover that nearly 90 percent of such meteorites are stony in composition—a value close to the volume proportions of the Earth's mantle. The slightly larger ratio of core material on our planet (16 percent) relative to iron meteorite abundances (about 10 percent) suggests a more complete metal-silicate separation in the more massive Earth compared with accumulation and planetesimal chemical differentiation in the asteroid belt. The similarities of meteoritic and terrestrial materials, and their comparable relative proportions, provide yet another compelling line of evidence supporting our previous inferences regarding the internal constitution of the Earth.

Geochemical and Heat-Flow Data

Even if the Earth had begun to accrete cold due to the gravitational collapse of a locally dense portion of the solar nebula, successive impacts by late-arriving planetesimals would have tended increasingly to heat the growing planet. This is a consequence of the fact that energy released on collision (converted to mechanical and thermal energy) is proportional to the masses of the colliding bodies. Inasmuch as radioactive decay also liberates thermal energy, long-lived and, especially, short-lived unstable radioactive isotopes would have contributed to this early-stage heating. Finally, the melting at depth of particles of iron, coalescence of droplets, and descent of molten metal down the gravity gradient to form an enlarging core would release large amounts of heat—apparently enough to raise the temperature of the entire Earth an average of about 2000 degrees Celsius. All these phenomena would have acted in concert to produce high temperatures and a strong thermal gradient in the primordial Earth, with the interior much hotter than near-surface layers of rock. Heat must have moved toward the cooling surface, as it does today, by a variety of mechanisms, chiefly radiative transfer, bodily flow of material (convection), and thermal conduction. Incipient melting and degassing would have resulted in the upward migration of low-temperature melting constituents, fluids, and large-radius elements insoluble in dense minerals found at great depths.

These processes would have enriched the outermost shell of the Earth (the crust) in fusible, large-radius, and volatile constituents such as we see concentrated in near-surface rocks today. Most of the heat-producing, unstable radioactive elements—those of potassium, uranium, and thorium—appear to be localized in this peripheral rind of the solid Earth, judging from the large proportion of observed heat flow that is generated by radioactive decay in the crust. If similar quantities of such radioactive elements were sequestered in the mantle or core, amounts of heat vastly greater than those measured would be emerging at the surface today. So we know that both mantle and core are nearly devoid of radioactive elements. On the other hand, provided most of the heat-producing, unstable isotopes are concentrated in the crust (as evidently is the case), the overall computed terrestrial abundances of potassium, uranium, and thorium turn out to be nearly identical to the proportions of these constituents in meteorites particularly, and the solar system in general.

A highly schematic temperature-pressure diagram showing the modern thermal gradient of the Earth (increase of temperature with depth) and the overall planetary differentiation into core and mantle is presented in figure 1.10. Also illustrated are idealized curves for the melting of metallic iron and the incipient fusion of mantle material. Clearly, the Earth's geothermal gradient coincides with the partial melting curve for magnesium-rich silicates only in the upper 100–220+ kilometers of the mantle; at both substantially greater and shallower depths, the mantle should be completely solid. The inner-outer core boundary must be a common point on both the iron (plus minor nickel) melting curve and the Earth's geothermal gradient. We know this by virtue of the fact that at the inner-outer core boundary, liquid and solid iron-nickel alloy are stable together.

Thus we see that the observed heat flow and geochemical concentrations of fusible, volatile, and large-radius elements near the surface of the Earth are compatible with its gross structure and inferred chemical differentiation. A more detailed, but nevertheless speculative, scenario for the evolution of the Earth will be presented in chapter 7.

Magnetic Data

The general nature of the global magnetic field is illustrated in figure 1.11. It is clearly dipolar (i.e., it possesses two poles), but at present the magnetic and geographic poles (the latter constituting the Earth's spin

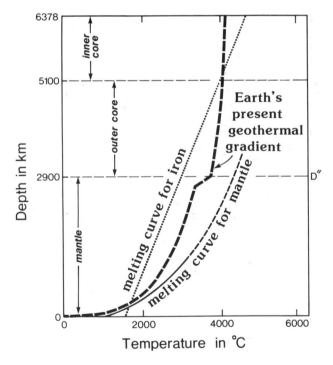

FIGURE 1.10. Highly diagrammatic temperature-depth relations in the present-day Earth. In reality, the melting curves for iron and for the onset of melting in the mantle are not smooth, but the general trends are approximately as illustrated. Different estimates of the temperature at the inner core–outer core boundary range from about 4000 degrees to nearly 7000 degrees Celsius. The temperature decrease upward through the D″ zone bounding the base of the lower mantle may be greater than shown.

axis) do not coincide exactly. Moreover, the field lines (directions of the magnetic force field) are observed to shift slowly with time. The origin of this magnetic field is not completely understood, but the most widely accepted theory holds that the Earth's outer, liquid core acts as a dynamo. Fluid flow within this electrically conducting material, and differential motion between the outer core and the base of the mantle, are thought to induce the surrounding magnetic field, which interacts in turn with the external solar wind. This dynamo evidently is governed somehow by the Earth's rotation, so the time-average magnetic pole positions probably coincide with the spin axis of the planet. In support of the hypothesis is the observation that all planets in our solar system that possess a dense, metallic core also appear to be enveloped by a magnetic field.

DYNAMIC NATURE OF THE EARTH

No one can study a modern relief map of the continents and ocean basins (figure 1.6) without being struck by the remarkable, congruent disposition of the continental borders framing the Atlantic Ocean; equally

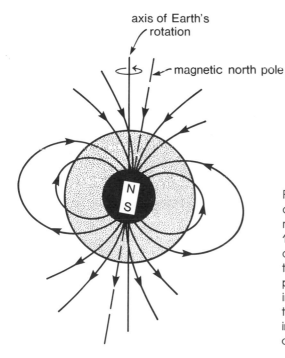

FIGURE 1.11. The magnetic field of the Earth (after Leet, Kauffman, and Judson 1987, figure 11.7). The inclination of the lines of magnetic force are proportional to latitude. We can approximate the field by imagining a bar magnet passing through the Earth's center and inclined 11.5 degrees to the spin axis (the pole of rotation).

astonishing is the existence along the middle of this ocean basin of a longitudinal, submerged mountain range that faithfully reflects the outlines of the continental margins! Other, equally stupendous submarine mountain chains, or rises, characterize the Circumantarctic, Indian, and eastern Pacific oceans. They exceed the length and breadth and rival the topographic ruggedness, of on-land mountain belts. Perhaps just as impressive first-order features on submerged portions of the Earth's rocky surface are the oceanic deeps, or trenches, that fringe the island arcs of Indonesia, the western Pacific, Japan, the Kamchatka Peninsula, the Aleutians, Central and South America, the Scotia arc, and the Lesser Antilles. The most prominent mountain systems on the continental crust completely encircle the Pacific, and link up with the transverse Alpine-Mediterranean-Himalayan-Indonesian zone; most segments are just landward from the oceanic trenches. What interplay of forces, so obviously lacking on our planetary neighbors, has resulted in this diversity and richness of topographic/hydrographic expression on Earth?

Geologists who have studied the continental crust learned early in the nineteenth century that strata of great antiquity are exposed on dry land. These old units are being eroded and carried away piecemeal towards the sea. Erosion is principally a consequence of a surficial heat-transfer process. Energy from the Sun preferentially warms the equatorial oceans, produces moisture-laden clouds in the atmosphere and currents in the hydrosphere. These processes ultimately effect the transport of H_2O via weather systems to the land, where cooling and precipitation result in the flow of water—ultimately back to the sea with entrained crustal debris. And yet, the Earth is not a flat, monotonous plain; our planet has always had mountains. Evidently, processes hidden from our view are at work deep within the Earth, slowly, inexorably thrusting sections of the continental crust to higher elevations. This inner turmoil is the planetary mass circulation fueled by internal heat that was alluded to earlier. The crustal manifestation of the bodily overturn is the mountain-building process, also called orogeny. This activity deforms pre-existing rocks, is commonly accompanied by volcanism, and generates lofty mountain ranges. It is the process that somehow accounts for the present crustal structure of the Earth and its surficial topographic/hydrographic manifestations.

In part, uplifted segments of the sialic crust are a reflection of the phenomenon of gravitative equilibrium in the Earth's crust, termed isostasy, as will be discussed. Briefly, measurements of the Earth's gravitational field have shown that low-density, topographically-high continental crust is underlain by a large sialic root that extends downward into the uppermost mantle much like the subsurface portion of an iceberg floating in the sea. The mountain height and elevated excess of continental mass is thereby gravitatively compensated through a displacement of denser mantle by light sialic material at depth, resulting in an equivalent mass deficiency. Consequently, the measured gravity field shows neither a marked positive nor negative anomaly. (Anomalies are departures from the expected field.) Similarly, low-lying continental crust is relatively thin, and the distance to the Moho is less than that under mountain belts. Therefore, different portions of the sialic crust range from about twenty-five to more than sixty kilometers in thickness. The denser oceanic crust is even thinner than the thinnest continental crust (the M discontinuity is only about five kilometers below the sea floor), and hence ocean crust is confined to deep basins.

Two different mechanisms of gravitative compensation are thus recognized: (1) where crusts of the same aggregate densities are involved, the thicker crust will rise to higher elevations than the thinner; (2) where crusts of similar thicknesses but contrasting densities are involved, the lower-density crust will rise to higher elevations than the denser. This relative buoyancy is a reflection of Archimedes' principle (wherein an immersed body apparently loses weight equal to the amount of liquid displaced), as illustrated schematically in figure 1.12. Because of readjustments in the asthenosphere, mass balance (i.e., isostatic equilibrium), is closely approached during loading and/or erosion of the Earth's outer rind. Some vertical crustal movements are thus a direct consequence of isostatic compensation.

Although we are far from a complete understanding, the form and diversity of the Earth's crust seems to be a reflection of both vertical and horizontal flow within the mantle, and the consequent differential movement of enormous lithospheric plates. This is the subject of the next chapter. Suffice it to state here that the chemical heterogeneities and topographic relief expressed in the continents, ocean basins, mountains, and oceanic trenches are reflections of the dynamic, internally circulating nature of our evolving planet, as well as surficial modifications due to erosion, transportation, and deposition. This evolutionary mobility, in turn, is a result of the Earth's attempt to redistribute more evenly buried heat as well as incident solar energy.

☰ METEORITE IMPACTS AND EARTH HISTORY

Thus far we have referred to the degradation of the Earth's surface as if reworking were due solely to the action of terrestrial agents of erosion (mass slumpage, and transportation of earth materials in moving geologic media such as wind, glaciers, and, most importantly, water), and this is true today. However, as alluded to earlier in the chapter, the primordial accretion of the Earth from planetesimals during collapse of the solar nebula resulted in rapid initial growth of our planet, followed by a gradual tail-off in the number of impacting meteoritic late arrivals as time progressed. The same story probably holds throughout the solar system as sweep-up of the planetesimals by the enlarging planets took place.

Once the Earth had achieved a considerable size and mass, collision with a moderate-sized asteroid—say, ten kilometers in diameter—would drastically affect the terrestrial environment during and immediately

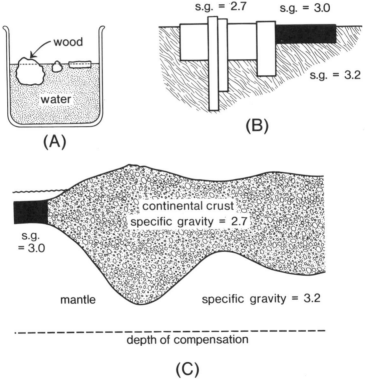

FIGURE 1.12. Archimedes' principle and isostatic balance of Earth's crust: (A) blocks of rather dense wood of uniform density (specific gravity equals 0.8) floating in a glass of water; (B) blocks of sial (specific gravity equals 2.7) and sima (specific gravity equals 3.0) "floating" in a mantle substrate (specific gravity equals 3.2); (C) rugged continental topography and corresponding low-density sialic roots. Note that the oceanic crust rides even lower than the thin continental crust. The mass of any column of unit cross section down to an arbitrarily deep level of compensation—say, the top of the asthenosphere—is everywhere the same. All three examples show the principle of isostasy.

after impact. The tremendous amount of energy released when a meteorite of this size hit the Earth would cause excavation of a major crater, wholesale partial melting of the upper mantle and crustal veneer, regional vaporization of seawater, and the injection of great quantities of particulate matter into the atmosphere. This dense dust cloud would envelop the Earth, block out sunlight, and raise the planetary reflectivity for months, years, or perhaps even decades, causing a pronounced global lowering of surface temperatures, and the onset of a severe climatic crisis.

The actuality of the occurrence of specific impact events and the magnitude of the effects are still being debated, but it seems likely that a major meteorite collision could profoundly and adversely influence the Earth's biosphere, as well as its physical environment. One such impact is hypothesized by some earth scientists to have taken place approximately 66 million years ago. This event is held responsible for worldwide mass extinctions among many species of plants and animals, including a most famous group of organisms—the dinosaurs.

Although detailed treatment of terrestrial biospheric history is beyond the scope of this book, less (internally) dynamic planetary neighbors including the Moon, Mars, and Mercury bear indisputable evidence of abundant meteoritic collisions (figure 1.2). It is certain therefore that, especially early in the life of our planet, the Earth too was subjected to such bombardment. The course of biological evolution almost certainly has been—and will continue to be—capriciously, and adversely, influenced by the orbits of Earth-approaching asteroids.

2
PLATE TECTONICS: ORIGIN AND DRIFT OF CONTINENTS AND OCEAN BASINS

Geologists have studied the structures of the continents for nearly two centuries, and much is known concerning the diverse origins of the uppermost part of the sialic crust, and of the constituent rocks and minerals as well. Until recently, however, the sima-floored ocean basins were more poorly understood. Within the past thirty years, the results of marine research have spectacularly illuminated the bathymetry (measurement derived from sounding ocean depths), structure, and physicochemical nature of the oceanic crust. As a result, we have a far better appreciation of the manner in which various segments of the Earth's outer rind have been produced, have evolved with time, and have been altered or destroyed. A startling product of this work has been the realization that portions of the Earth's mantle are slowly circulating, or convecting, producing new oceanic crust along submarine ridge (oceanic rise) systems, and both adding to, and deforming, the continental crust in the vicinity of seismically and volcanically active continental margins and island arcs. Oceanic rises are sited over upwelling mantle sheets or columns known as plumes, whereas mantle currents are descending beneath active continental margins and island arcs. Just what propels this asthenospheric flow is uncertain, but that some sort of circulation is taking place is beyond dispute.

It is now known that both continental and oceanic crust form only the uppermost segments of great lithospheric plates, whose differential motions are a reflection of the mantle convection process. The continents of the Eastern and Western Hemispheres are presently drifting apart from one another in the vicinity of the Atlantic Ocean and have been diverging for more than 120–190 million years; locally, sialic fragments came together in the past and others are currently on a collision course, especially around the Pacific Rim. The idea of continental drift was proposed more than seventy years ago by the German meteorologist Alfred Wegener. Nevertheless, his repeatedly and earnestly advocated concept was almost universally rejected by the earth science community and was ignored for nearly fifty years. Then from 1963 to 1968, virtually all geologists and geophysicists abruptly changed their minds about continental drift. Why did such a reversal in thought take place so swiftly and so completely? Let us look at what happened.

▬ HISTORICAL DEVELOPMENT OF THE CONTINENTAL DRIFT THEORY
Wegener's Scientific Predecessors

Although we will not dwell on the historical development of the earth sciences, Wegener's hypothesis has profound implications for an appreciation of the origin and evolution of the outer portions of our planet. Furthermore, the controversy provides an unsettling insight into the operation of the scientific method, and the manner in which models of Earth processes have changed with time. Thus it is worthwhile to note some of the antecedant observations that led to the unconventional idea of drifting continents, as well as arguments concerning the concept itself.

About the year 1620, Francis Bacon called attention to the general conformity of the outlines of the eastern coast of South America and the western coast of Africa. Obviously, such a geometric relationship could not have been deduced until the navigators of the Renaissance—the Age of Exploration—charted the seas and land masses sufficiently well for accurate maps to be produced by cartographers of the time. Bacon did not pursue the implication of his observation, but, in retrospect, it clearly would have to be a remarkable situation if the fit of these continental margins were merely a coincidence. Scientists, like detectives, tend to feel very uneasy about explanations of phenomena that appeal only to coincidence.

In 1858, Antonio Snider-Pellegrini concluded that South America and Africa once must have been joined together, and suggested that their disjuncture resulted from the biblical flood. By this time, of course, most geologists were convinced of the great antiquity of the Earth, based on observations of the almost infinitely slow processes (erosion, deposition, etc.) that shape the surficial portions of the crust. Charles Darwin's theory of evolution supported this view through documentation of the extraordinarily slow progress of biological changes recorded in fossiliferous rocks. Thus earth scientists were convinced that the Earth was many orders of magnitude older than the value arrived at by Archibishop Ussher based on his studies of the Old Testament, and they were equally unreceptive to so-called geologic hypotheses based on the Noachian flood.

Twenty-one years later, Osmund Fisher accepted George Darwin's imaginative but rather rash speculation that the Moon represented continental crust wrenched out of a surficial segment of the Earth during Cretaceous time (see figure 7.2 for the geologic time scale), leaving behind the great Pacific Basin. Fisher called on this cataclysmic event to initiate drift among the surviving continents due to a mass-deficiency-driven crustal instability. The absence of profound geologic and topographic changes in both the remaining continents, and in life, both on the land and in the seas, as a direct consequence of the alleged formation of the Moon in this manner cast important doubt upon this exciting speculation.

Finally, in 1910, one of Wegener's contemporaries, F. B. Taylor, postulated that roughly east-west-oriented arcuate (curvilinear in map view) mountain chains, such as the Alpine-Mediterranean-Himalayan-Indonesian belt, reflect compressional features due to the rotationally impelled (centrifugal) movement of continental masses away from the Earth's geographic poles and towards the Equator. Taylor seems not to have worried excessively about more nearly north-south ranges, such as the Rocky Mountains, Andes, Urals, Appalachians, and Caledonides, among others.

Wegener's Bold Hypothesis and the Ensuing Controversy

Alfred Wegener (1880–1930) was a German meteorologist and geographer who taught in Gratz, Austria, and conducted much of his research in Greenland. It has been speculated that his ideas regarding the breakup and drift of continents might have been inspired by observation of the rifting and drifting of pack ice floating in the Arctic Sea. Wegener never alluded to this phenomenon in his writings, however. In any case, he published and refined his hypothesis on the origin of continents and oceans repeatedly (1912, 1915, 1924, 1929). The basic idea was that the sialic continents float on a denser but weaker simatic substrate, the oceanic crust, much as ice, immersed in water, floats. Like icebergs, the continents maintain structural continuity, but episodically have been fractured and subsequently drift apart. Wegener came to a tragic and untimely end on the Greenland ice cap and did not live to see his ideas vindicated in modified form during the 1960s.

Wegener was impressed, as was Bacon three centuries earlier, by the remarkable congruence across the Atlantic of the eastern- and western-hemispheric coastlines. (A modern computer fit of these continents is illustrated in figure 2.1.) He logically concluded that such a geometry could only have resulted from an irregular rupture and displacement—like pieces of a gigantic jigsaw puzzle—of portions of a once-contiguous land mass. In hindsight, the geometric argument ought to have been sufficiently compelling, but it was not because the fit was less than perfect. Unfortunately, Wegener chose the shorelines to represent the continental margins, whereas he should have selected the outer limit of the continental shelves, or midway down the continental slope, for these more accurately represent the actual continental crust edge.

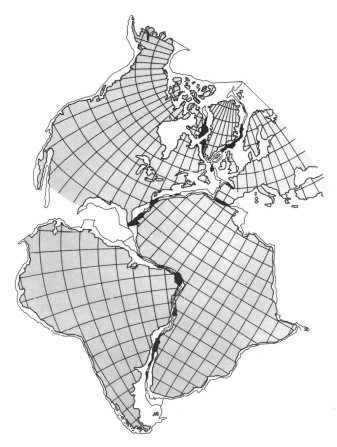

FIGURE 2.1. The fit of continental land masses across the Atlantic Ocean, generalized after Bullard, Everett, and Smith (1965). The gaps are shown in white, the overlaps (chiefly post-rifting sedimentary deltaic deposits) in black. Present-day longitudes and latitudes are superimposed on the sialic crust for reference.

Of course, the submerged extents of the continents were not accurately known in Wegener's time. The reasons the "pieces of the puzzle" can be restored closely but not exactly is a reflection of the existence of only narrow continental shelves, and of the fact that both interior and marginal zones of the continents are being deformed by mountain-building processes. Furthermore, erosion and deposition over the millions of years following the initiation of drifting have somewhat modified the edge geometries of the once-continuous continental fragments. Nevertheless, if you reexamine figure 1.6, you will be impressed with the congruence across the Atlantic. Note, too, that New Zealand and adjacent submerged plateaus fit readily into the east coast of Australia. Where, off the east coast of Africa, would you place the island of Madagascar? Even professionals argue about this one!

As shown in figure 2.2, Wegener employed the geometric-outline argument, and other evidence to be discussed below, to reconstruct the entire continental assembly as it apparently existed some 200–250 million years ago. This completely-joined-together land mass he called Pan-

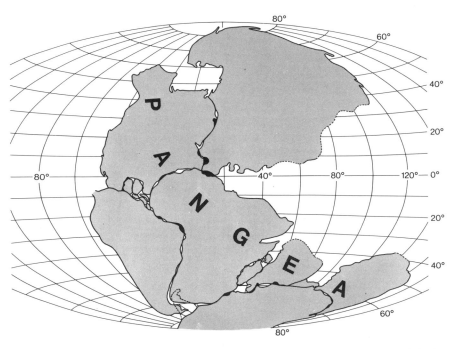

FIGURE 2.2. Wegener's original continental assembly, Pangea, appropriate for the end of the Paleozoic Era, about 250–200 million years ago (simplified after Dietz and Holden 1970).

gea. It consisted of two major segments (supercontinents): (1) North America plus Eurasia, or Laurasia; and (2) South America, Africa, India, Australia, and Antarctica, or Gondwana. Some portions of this supercontinental assembly, such as Australia plus Antarctica plus New Zealand, and Africa plus South America, seem to fit together readily, whereas ambiguity exists as to the locus of joining together other portions of the sialic crust. The following, somewhat controversial, lines of evidence led Wegener to his particular reconstruction.

Fossil plants and animals found in Permian and Triassic strata (roughly 205–285 million years old; for a brief discussion of the geologic time scale, see chapter 7 and figure 7.2) exhibit striking similarities in southern South America, South Africa, peninsular India, and parts of both Antarctica and Australia-plus-New Zealand. Geographic relations are illustrated in figure 2.3. The so-called *Glossopteris* flora (chiefly Permian and Triassic) is confined to the more southerly (in present coordinates) parts of the Gondwana supercontinent, where it constitutes a distinctive, temperate-climate assemblage. Such fossil plants are totally absent from equivalent ecological environments represented by deposits of the same age in the Laurasian supercontinent, because they could not propagate, flourish, and be distributed through equatorial paleoclimatic zones (climatic zones of past ages).

The same situation applies to the distinctive *Therapsida*, reptiles that exhibit skeletal features transitional toward primitive mammals,

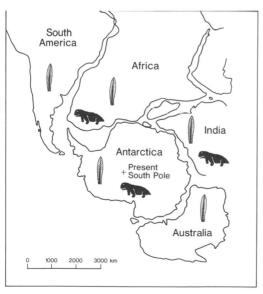

FIGURE 2.3. Fossil fauna and flora unique to the Gondwana supercontinent (schematic). Illustrated biota! now extinct, were known to make good pets.

and that are similarly confined to southern portions of Gondwana. Assuming positional permanency of continents and oceans, early paleontologists had evoked the ad hoc concept of land bridges—dry-land corridors, or chains of islands, now vanished beneath the sea—to account for the far-flung occurrences of these unique plants and animals. The embarrassment of postulating the later sinking of such buoyant sialic materials into the mantle without leaving a trace (which flies in the face of the principle of isostasy discussed in the preceding chapter), and the difficulty in explaining why the biota failed to propagate northward into Eurasia and North America, were neatly avoided by the continental drift hypothesis. During the controversy, however, paleontologists questioned the extent of faunal and floral similarity, and in general disputed Wegener's alternative explanation for the observed wide geographic dispersal of these distinctive plants and animals.

Bedded Permocarboniferous and Triassic sedimentary rocks (roughly 200–300 million years old) contain evidence of climates attending the times of deposition that seem strangely out of place with regard to their modern locations. For instance, bouldery gravels and clays indicative of continental glaciation, such as characterize polar and circumpolar regions, occur in present-day temperate and tropical latitudes in South Africa, Patagonia, central Australia, and peninsular India. Even more surprising are the equally old occurrences of abundant, thick coal beds and layers of salt, currently situated in high latitudes—e.g., Antarctica, Alaska, Spitzbergen, and East Germany. Such accumulations are forming today principally at near-equatorial latitudes in excessively moist (rain forest) and arid (desert) climates, respectively. Although drifting of the continents with respect to the Earth's spin axis readily accounts for the geographically linked changes in paleoclimatic zonation, many of the individual occurrences were disputed by "stabilists" as due to local, topographically induced climatic anomalies, whereas other deposits were said to be incorrectly described and interpreted.

Another point stressed by Wegener involved the origin of compressional mountain belts. These so-called orogenic chains contain rock units that show evidence of great crustal shortening, for they appear to have been folded and thrusted much like the rumpling of a rug. Wegener argued that such structures form at the leading edges of continental masses that are "plowing through" the ocean basins (refer again to figure 1.6). Examples include the Circumpacific mountain chains (Andes, Aleutians, Kamchatka Peninsula, Indonesia, the islands of Japan, etc.); the great Himalayas and the Tibetan Plateau were regarded as a product of the collision of the Indian Peninsula with the "soft underbelly" of Asia. The unconvinced responded, with some reason, that if the sialic continents were strong, integral structures moving through the weaker, simatic oceanic crust, it should have been the oceanic crust, instead of the continental, that was deformed—and this is manifestly not the case. Moreover, Wegener's model failed to explain the continental interior locations of great mountain chains, including the Rockies, the Appalachians, and the Urals.

Another supportive observation advanced by Wegener concerned the fact that geologic trends (sedimentary basins, ancient recrystallized cores of the continents, distinctive linear igneous belts, etc.) are truncated at the margins of Africa, but are matched by similar forms and features in South America, Australia, and India. Some of these abruptly terminated geologic provinces are illustrated schematically in figure 2.4. (For a discussion of igneous, metamorphic, and sedimentary rock types, refer

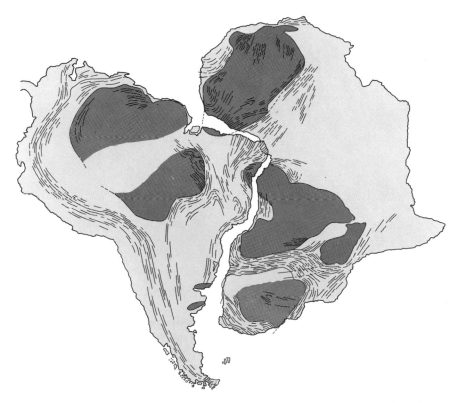

FIGURE 2.4. Geological belts truncated at the present margins of the South Atlantic Ocean. Dark areas are early Precambrian cratons, lineated patterns show the trends of late Precambrian and younger mountain belts (after Hurley 1968).

to chapter 4.) The trends of these belts are continuous when the segments of Gondwana are restored to their pre-drift configurations. Nonbelievers countered that such correlations were highly interpretative, and that analogous trends across the North Atlantic were even more speculative. They faced a serious problem, however, in the abrupt termination of geologic features at some continental margins and were obliged to erect the hypothesis of now-vanished borderlands that, like the problematic land bridges, disappeared mysteriously beneath the seas, isostasy notwithstanding.

The crucial argument, however, that delayed acceptance of the concept of continental drift, was that put forward by the British physicist Harold Jeffries. Wegener believed that the motions of the continents subsequent to the breakup of Pangea were largely toward the Equator as a consequence of the rotation-generated oblateness (polar flattening) of the Earth; the postulated westward component was supposed to be a response to tidal forces. However, laboratory measurements of rock strengths demonstrated the apparent rigidity of both oceanic and continental crust. Thus, Jeffries was able to show quite conclusively that the outer portions of the Earth were much too strong to allow motions of the type envisioned by Wegener. Furthermore, the onset of rifting and drift proposed by Wegener allegedly began about 190 million years ago in the proto–North Atlantic, while separation of South America and Africa took place about 120 million years ago: both at times, not of widespread mountain building and crustal unrest, but of near worldwide quiescence. On Jeffries' evidence, the geological unreasonability as well as the geophysical impossibility of Wegener's concept were emphasized by the antidrifters.

Yet now we know that Wegener was largely correct. What had we all missed? His bold concept was rejected chiefly because scientists could not explain how Wegener-style drift could be physically possible. Ironically, the mechanisms of continental drift are still being debated today, although we now agree that, without a doubt, the continents are drifting!

MODERN CONCEPTS OF SEA-FLOOR SPREADING AND PLATE TECTONICS

Even though the structure and constitution of the continental crust have been investigated systematically since the beginning of the last century, until recently we have had a most imperfect understanding of the mountain-building process. Nor had we a much more satisfying answer as to why the agents of erosion (mass slumpage, wind, moving ice, and, especially, running water) have failed to wear down the land to a featureless plain at, or below, sea level. After World War II, classified bathymetric data were released; more importantly, scientists put out to sea with new equipment and different kinds of information became available. These marine observations, employed in conjunction with the known terrestrial geology, have allowed a much fuller appreciation of processes at work on our planet. The new data are of several different types; among them must be included the varied topography of the ocean bottom (bathymetry), marine linear magnetic anomalies, belts of seismicity and the inferred existence of lithospheric plates, and sediment patterns and ages of deposition in the ocean basins.

Bathymetric Features of the Ocean Basins

The sea floor lies, on the average, about five kilometers below the water surface, but some of it is far from flat, as may be seen by examination of figure 1.6. The presence of oceanic deeps (trenches), that lie offshore from certain island arcs and are especially abundant in the western Pacific, had been known prior to 1940, but their linear extent and continuity had been insufficiently appreciated.

The most astonishing observation, however, was the discovery of globe-encircling oceanic ridges—or rises—beneath the seas. These are constructed predominantly of basalt (the dark, ferramagnesian lava of Hawaii) and its deep-seated equivalent, gabbro (see chapter 4 for descriptions of rock types). The Mid-Atlantic Ridge, illustrated in figure 2.5, lies equidistant from Europe-plus-Africa on the east, and the Americas on the west. Knowledge of its significance surely would have fueled Wegener's arguments, had the ridge been accurately delineated during his lifetime. The East Pacific Rise, which seems to terminate within the Gulf of California (but reappears for a short stretch off the coast of Oregon-plus-Washington-plus British Columbia), is in fact the northern extremity of another major ridge that extends southwestward through the Pacific Ocean, westward between Antarctica and Australia, and then curves northwestward through the Indian Ocean to end in the Gulf of Aden–Red Sea rift complex. A splay trends southwestward around South Africa, separating it from Antarctica, and on eastward to the Mid-Atlantic Ridge (refer to figure 1.6). Yet another ridge system, the northern extension of the Mid-Atlantic Ridge, transects the Arctic Ocean before losing its identity.

Other bathymetric features in the ocean basins include oceanic plateaus, isolated, dominantly volcanic islands, linear island chains, and seamounts. These last are islands that are now submerged to considerable depths; many, those known as guyots, are flat-topped. Except for some subsea plateaus that may be fragmented blocks of sialic crust, all

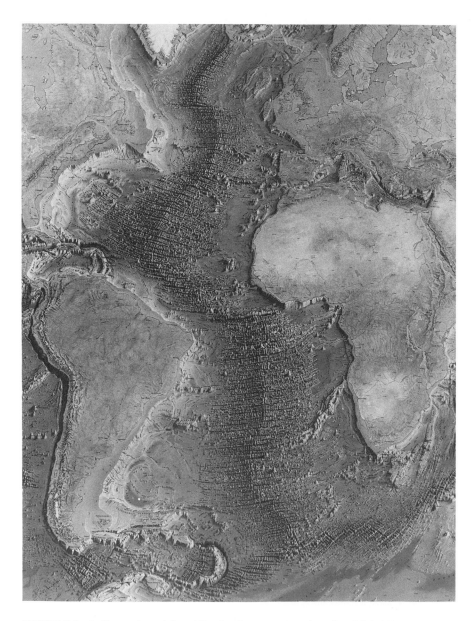

FIGURE 2.5. Bathymetry of the Atlantic Ocean, showing the Mid-Atlantic Ridge, the prominent submarine mountain chain located equidistant between eastern and western hemispheric continents (Heezen and Tharp 1977).

these features are simatic in composition. Oceanic fracture zones, which manifest themselves as submarine escarpments, are particularly obvious and numerous as ridge offsets (see figure 1.6), but such physiographic features also transect island chains, and a few apparently offset segments of island-arc-and-trench complexes.

The midocean ridges and rises are regarded as the near-surface expression of almost imperceptibly ascending mantle currents. Whether this upwelling is due to mantle return flow accompanying plate sinking elsewhere, or is a consequence of thermal anomalies and deeply buried mantle buoyant masses, is not known. Upward-trending velocities of mantle materials are on the order of a few centimeters per year. On approach toward the sea floor, the rising asthenosphere undergoes decompression partial melting (incipient melting in response to lowered pressure at constant temperature; see chapter 4) to produce modest amounts of iron-and-magnesium-rich basaltic magma (molten rock). These liquids are less dense, more buoyant than the solid parental mantle material, and move more rapidly upward toward the seawater, where they solidify to form a simatic carapace, the oceanic crust, over the residual solid, but plastic, mantle. As the complex ascends and cools, it bifurcates, rifts, and moves at right angles in both directions away from the ridge axis; in departing, it is succeeded by a rising complex of basaltic melts and underlying softened mantle material. Oceanic ridges are termed spreading centers, because the lithospheric slabs or plates surmounting the mantle currents move orthogonally (at right angles) away from the axis. This type of plate junction is known as a divergent boundary; a schematic view of its geometry is shown in figure 2.6. The process is termed sea-floor spreading; it produces lithosphere from athenosphere.

On cooling, the lithosphere thickens with time as it migrates away from the ridge or spreading axis. This is a consequence of the fact that, as heat is lost from the upper mantle, temperatures high enough to result in partial melting are confined to greater and greater depths. The lithosphere-asthenosphere boundary (solid mantle above, incipiently molten mantle below), which is very close to the sea bottom beneath the ridge, therefore descends to greater depths away from the spreading axis. As the lithosphere cools and thickens with the passage of time, its aggregate density increases; accordingly, it rides on the asthenosphere at lower and lower elevations below sea level due to isostatic compensation. Unlike low-specific-gravity sialic continental masses floating on a denser substrate, the lithosphere has a higher specific gravity than the underlying asthenosphere. The situation involving cooler oceanic crust—capped lithosphere surmounting hotter asthenosphere is therefore gravitatively unstable: the lithosphere will sink where geometrically possible.

Eventually, due to sea-floor spreading, the material making up a lithospheric plate may reach a convergent boundary with another plate. One slab must descend into the mantle in order to conserve volume. This process, called subduction, is favored by the density instability described above; a diagrammatic sketch is presented as figure 2.7. Usually, although not invariably, the oceanic crust–capped plate, being denser, thus negatively buoyant, slides down into the mantle. Rarely, a continental crust–capped lithospheric plate may begin to sink, but this is usually a consequence of the prior descent of oceanic crust–capped lithosphere, leading a microcontinental fragment into the subduction zone. In any case, the locus of the downturn, where bending of the downgoing slab is maximized, is marked by a bathymetric low, or trench.

Not all the volcanism in the ocean basins occurs along the spreading centers. Isolated activity builds islands and island chains within plates as well. This igneous phenomenon is due to thermal anomalies, or hot spots, thought to be located at considerable depth in the upper mantle. Volumetrically, the amounts of melt generated are not important compared to that produced at the midocean ridges. Two points should be mentioned, however.

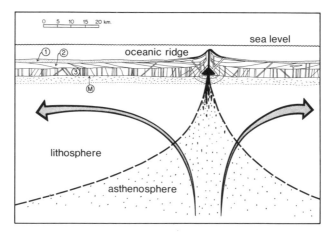

FIGURE 2.6. Diagrammatic cross section of a midocean ridge or divergent plate boundary. Mantle flow lines lie in the plane of the paper. Basaltic magma is illustrated as black blobs. Layers 1, 2, and 3 are deep-sea sediments, basaltic flows, gabbroic and diabasic intrusives, respectively (see chapter 4). Directly below the Mohorovicic discontinuity, M, lies a zone of settled crystals, the bulk composition of which is similar to the underlying mantle. The base of the lithosphere, indicated by dashed lines, is the surface along which the solid upper mantle softens downward apparently due to incipient melting at higher temperatures.

First, midplate volcanism results in the construction of seamounts, some of which rise above wave base (the depth in the ocean—typically a few meters below sea level—beneath which surface wave action is absent). After cessation of eruptive activity, moving water, and other erosive agents plane off portions of the volcanic edifice that project above wave base—doing so at a geologically rapid rate. Due to isostasy, such flat-topped mountains gradually sink beneath the sea as the lithosphere ages, cools, and thickens, resulting in deeply submerged, erosion-truncated volcanoes. In tropical regions, fringing coral reefs tend to grow upward at about the same rate as the island subsides, eventually forming ring-shaped atolls; this mechanism was deduced by Charles Darwin more than 130 years ago. Flat-topped seamounts (guyots) and atoll structures are illustrated in figure 2.8.

The second point worth stressing is that many of these midplate volcanic centers may reflect activity associated with some sort of high-temperature region situated deep in the upper mantle. The origin of

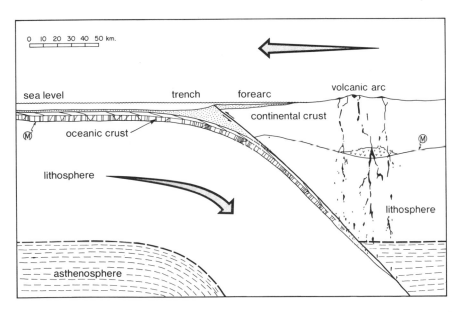

FIGURE 2.7. Diagrammatic cross section of an oceanic trench and island arc/continental margin, or convergent plate boundary. A large component of mantle flow lies in the plane of the paper. Both basaltic and andesitic magmas (see chapter 4) are shown as black blobs. The base of the lithosphere, shown by dashed lines, is the region where the more rigid mantle above becomes more plastic downward, probably as a consequence of incipient melting at higher temperatures.

these hot spots is not well understood; scientists have speculated that they are the sites of large meteorite impacts, whereas others regard them as manifestations of deeply buried mantle heat sources. If a stationary thermal anomaly within the mantle persists over a long interval of geologic time, the passage of oceanic crust–capped lithosphere over the hot spot will result in the construction of a string of islands; members of these island chains are progressively older in the "downstream" direction. That is, a volcano is built on the moving lithospheric plate and is gradually displaced until it no longer is vertically above the mantle thermal anomaly and volcanic plumbing system. Extrusive activity ceases therefore. A new volcano is constructed directly over the hot spot magma source, is displaced in turn, and subsequently is shut off. The bearing of the linear island chain reflects the plate motion vector relative to the apparently immobile deep upper mantle. A very good example of this phenomenon is the Emperor-Hawaiian chain, as shown in figure 1.6. Note the deflection in trend from west-northwest to north-northwest about midway along the chain (near the island of Diakakuji); here volcanic activity occurred about 40 million years ago. Ask yourself what direction the Pacific plate was moving prior to this time; you may check your answer by studying figure 5.4. Twenty-five years ago no one knew what you have probably just surmised.

Oceanic fracture zones appear to reflect the spatial requirement that lithospheric plates move orthogonally (at right angles) away from the spreading axis so as to allow neither plate gaps nor mass excess. Even if

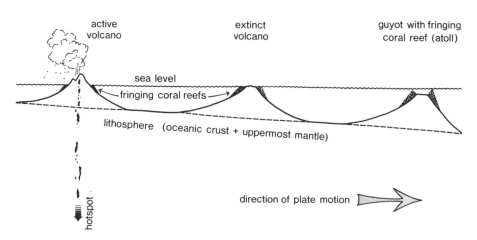

FIGURE 2.8. Schematic representation of the waxing and waning of midplate volcanic islands, flat-topped seamounts (guyots), and coral atolls. Ascending melts, which feeds volacanic activity, are shown as black blobs.

the original geometric form of an upwelling current initiating a divergent plate junction were curvilinear on the Earth's surface, the requirements of volume conservation would still demand straight, rectilinear ridge segments, as illustrated in figure 2.9. This style of motion is reflected in the observed disposition of magnetic anomaly patterns, variations in the observed terrestrial magnetic field we will describe later on in the discussion. You may think of each of the ridge segments as the axis of two back-to-back conveyor belts, the strips of sea floor as the moving belts. The oceanic fracture zones, or conservative plate boundaries (that is, neither plate creation nor plate destruction boundaries), are vertical planar discontinuities that offset ridge crest segments and are parallel to lithospheric and possibly asthenospheric flow lines. These offset planes are termed transform faults. Faults are fractures that have experienced differential movement of the opposite walls. In general, the surface expression of oceanic fracture zones may be described as small circles on a sphere, analogous to lines of latitude on the globe, whereas a ridge approximates an arc of a great circle, analogous to the Equator and to lines of longitude on the globe. Where transforms cut across a

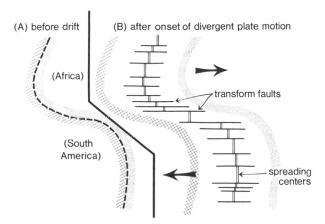

FIGURE 2.9. Hypothetical initiation of the Mid-Atlantic Ridge above an upwelling (asthenospheric) mantle current, the surface expression of which is assumed to have been an irregular, curvilinear feature. Note that parallel flow in the mantle and oceanic crust results in straight spreading-center segments offset by right-angle fracture zones, or transform faults (Mercator projection): (A) prior to spreading; (B) after a modest amount of divergent plate motion. Whether the plates are pulled apart by descent of a dense slab in the subduction zone, slide off the hydrographically elevated ridge, or move passively due to flow in the underlying asthenosphere, the lithospheric motions and return flow in the ductile mantle are the same.

portion of the continents, we call such features strike-slip faults, because the movement is parallel with the linear extension—or strike—of the fracture. Examples include the Anatolian fault of northern Turkey, the Alpine fault of South Island, New Zealand, the Median Tectonic Line of western Japan, and the San Andreas fault of western California.

Although we have spoken of plate motions as if they were linear translations, obviously the lithospheric plates are curvilinear segments (like pieces of the rind of an orange or, more properly to scale, segments of the skin of an apple) surmounting a spherical Earth. Plate motions therefore may be described as rotations about axes through the center of the globe; of course, the poles of such rotations need not—and in general do not—coincide with the Earth's spin axis (the geographic poles).

Marine Magnetic Anomaly Data

As we have seen, deep sounding of the sea floor by oceanographic surveys has greatly improved our understanding of submarine topography. Concurrent with these investigations, the Earth's near-surface magnetic field was measured by shipboard geophysicists. The local magnetic field exhibits variations in intensity—magnetic anomalies—that are superimposed on, and deviate from, the regional magnetic field. As illustrated in figure 2.10, such anomalies are disposed symmetrically about a midoceanic ridge as alternating bands of greater and lesser intensity than the expected magnetic field (positive and negative anomalies, respectively). Clearly, these "zebra stripes" must be related to some process taking place in the vicinity of the spreading system. What is their

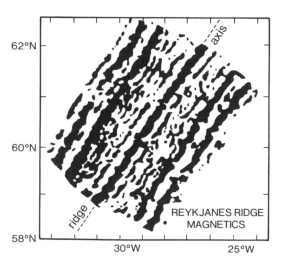

FIGURE 2.10. Magnetic anomaly pattern for a small portion of the Mid-Atlantic (Reykjanes) Ridge southwest of Iceland (after Heirtzler, LePichon, and Baron 1966). Positive anomalies are shown in black, negative anomalies in white. The bilaterally symmetrical striping continues both northeast and southwest, parallel to the extension of the ridge axis.

significance? To understand this phenomenon, we must discuss the concept of paleomagnetism.

About thirty years ago it was recognized that the Earth's dipolar magnetic field reverses episodically (at irregular intervals). This is called a magnetic reversal and means that the north-seeking end of a compass needle would point south during a period of reversed magnetic polarity. Of course, at present we are living in a time of normal magnetic polarity. But how can we be sure that the Earth's magnetic field has switched polarity every so often during the course of geologic time?

Rocks laid down at or near the surface, such as lava flows, volcanic ash layers, marine sediments, and lake beds, contain minor concentrations of magnetic minerals. Magnetic domains within these minerals, or the individual grains themselves, become aligned as a consequence of the Earth's external (total) field as it existed during formation of the rocks (see figure 1.11). When measured in the laboratory, such specimens thereby display weak remanent (remnant) magnetism that was imposed in response to—and coincident with—the external field extant during formation of the rock. Both the azimuth (magnetic north direction) and inclination from the horizontal (or dip) of the magnetic field lines at the time of origin can be recovered if the initial orientation during formation of the collected rock is known. The magnetic azimuth indicates normal or reversed polarity and the direction of the magnetic paleopole. The magnetic dip provides the magnetic latitude of the rock at the time of origin; in other words, the paleolatitude (see figure 1.11). When the remanent magnetism of a sufficiently thick sequence of layered rocks (youngest at the top, oldest at the bottom) is measured, it is found that groups of rock units possessing normal polarity are interstratified with layers characterized by reversed polarity. Moreover, where good fossil and/or radiometric data (see chapter 7) allow the precise assignment of ages to such a series, analogous stacks of layered rocks around the world can be shown to exhibit exactly the same time-versus-magnetic-polarity sequence. The geomagnetic time scale for relatively young rocks is illustrated in figure 2.11. Evidently the Earth's magnetic field undergoes polarity changes on an irregular frequency, but always maintains its dipolar character.

Thus, in 1963, when the widespread magnetic anomaly patterns of the ocean basins began to be recognized, it was concluded that the anomaly stripes were due to the remanent magnetism of the oceanic crust superimposed on the regional field: positive in cases where the basaltic lavas have normal polarity, negative where they congealed during a time interval characterized by reversed polarity. It follows that the crust at the ridge axis is youngest (we know that it is being formed there today), and that as one moves laterally away from the spreading center, progressively older basalts are encountered, accounting for the linear pattern of alternating positive and negative anomalies. A midoceanic ridge system, in effect, acts as a pair of back-to-back magnetic tape recorders, with the rocks faithfully reproducing a two-note sonata as the Earth's magnetic poles reverse from time to time. The rate of horizontal transport is on the order of several centimeters per year, judging from application of the geomagnetic time scale to the observed magnetic anomaly pattern. When this relationship was realized, sea-floor spreading and the opening of the Atlantic were quickly accepted, and continental drift had been proved beyond reasonable doubt.

Seismicity and Lithospheric Plates

We have seen that the bathymetry and remnant magnetism of ocean basins such as the Atlantic are compatible with a model of rising currents in the solid but plastic mantle beneath midoceanic spreading systems, or divergent plate junctions, and lateral movement of the lithospheric plates at right angles to the ridge axis, somewhat like two opposed conveyer belts, as previously alluded. Convergent plate junctions occur where ocean basins such as the Pacific are being constricted around

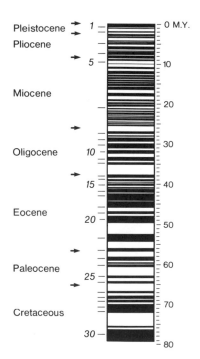

FIGURE 2.11. The geomagnetic time scale for the past 80 million years, showing worldwide intervals of normal (black) and reversed (white) magnetic polarity, as deduced from remnant magnetic measurements in well-dated rocks (after Heirtzler et al. 1968). Geologic periods and epochs, anomaly numbers, and age in millions of years, are indicated.

their margins, and one lithospheric slab sinks beneath another. The study of seismic wave transmission velocities and relative attenuation provided the first crucial evidence for the now universally accepted plate-tectonic model: seismic data unambiguously demonstrated the existence of stronger, more rigid lithosphere surmounting weaker, more plastic asthenosphere. What does seismicity tell us about the nature and relative motions of these lithospheric plates? Quite a lot, as it turns out.

Intense earthquake activity is virtually confined to the lithosphere, which—on the rapid application of stress—behaves in brittle fashion compared with the more ductile asthenospheric substrate. Moreover, within-plate seismicity is much less intense than along the slab margins, where frictional energy release reflects differential plate motions; hence, concentrated activity tends to outline present-day lithospheric boundaries. Examination of a world seismic map such as figure 2.12 provides us with an appreciation for the sizes and configurations of the plates. Incidently, the surface projections directly above the earthquakes, or epicenters, are shown in figure 2.12, rather than the actual depth locations, or hypocenters. Seven major plates, are outlined by this seismicity, as is clear from figure 2.13: the American, Eurasian, African, Antarctic, Nazca, Pacific, and Australian. In addition, a number of smaller plates are evident, such as the Cocos-Farallón, Arabian, Caribbean, and Philippine. Note that the near-surface portions of some plates are dominantly oceanic (e.g., Pacific, Nazca, Philippine), with others mostly capped by continental crust (e.g., Arabian, Eurasian); but in general the upper portions of the lithospheric slabs consist of both simatic and sialic crust. Here, of coure, is where Wegener's hypothesis was incorrect; he had no way of knowing that continents and ocean basins were merely the uppermost parts of the moving plates. When you consider it, the existence of these immense lithospheric plates still seems astounding.

The boundaries of the lithospheric slabs are of three types as previously described: divergent, convergent, and conservative. Midoceanic spreading ridges, trench-subduction zones, and transform-fault fracture zones embody these three types, respectively. Where upwelling mantle currents approach the surface, high-temperature solid but incipiently melting material (asthenosphere) lies at shallow depths. Heat flow is thus high, and the rocks deform through ductile flow rather than by rupture. For this reason, seismicity associated with spreading centers and transform faults is confined to depths of less than about fifty kilometers. In contrast, beneath the trenches, which are regions of abnormally low heat flow, downgoing slabs descend to depths of 600–700 kilometers before finally losing their rigidity and lateral integrity, as outlined by earthquake hypocenter locations. This is why the divergent, exclusively shallow plate junctions shown in figure 2.12 are well-defined

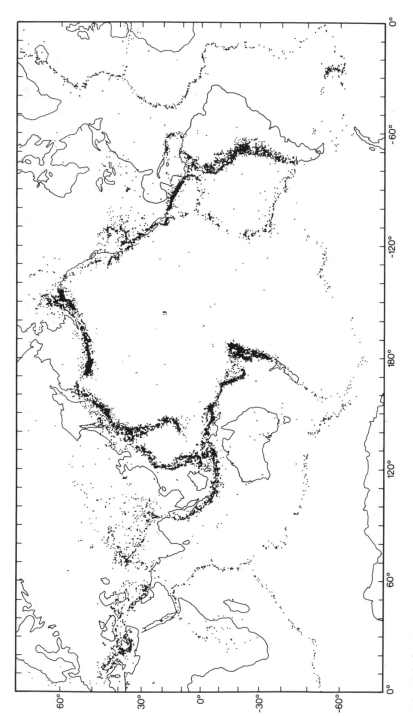

FIGURE 2.12. World seismic activity for the period 1961–67, after Barazangi and Dorman (1969). Dots mark earthquake epicenters.

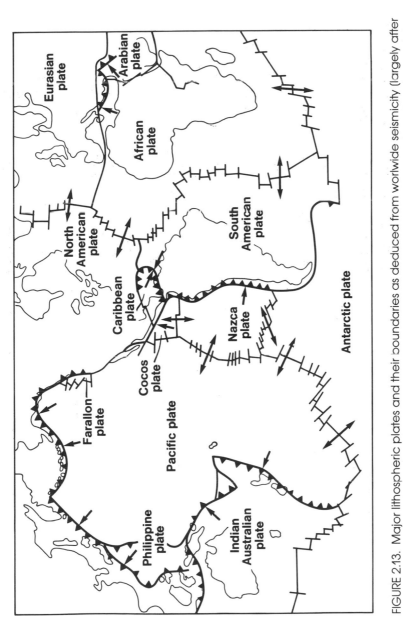

FIGURE 2.13. Major lithospheric plates and their boundaries as deduced from worlwide seismicity (largely after Dewey 1972, and Francheteau 1983). Convergent boundaries are shown with barbs on the upper plate, divergent boundaries by means of double arrows, and conservative boundaries with single, light lines.

by linear arrays of epicentral locations; in contrast, hypocenters from various portions of the inclined subduction zone characteristic of convergent plate junctions and deep-sinking slabs, when projected to the surface, are dispersed over a much broader region. Another indication that lithosphere is descending into the deep upper mantle along convergent plate boundaries is gained from the observation that local transmission velocities in the neighborhood of the subduction zone are quite high, and attenuations of seismic wave energy are unusually small for these mantle depths, suggesting the unusually deep presence of lithosphere.

The third type of plate boundary is the so-called conservative or transform lithospheric plate junction. Here two plates slide past one another in parallel without either diverging or converging. Although continental transforms (e.g., the San Andreas fault in California, the Alpine fault in South Island, New Zealand) are well known, the most abundant and clearest examples are found among submarine fracture zones offsetting segments of the midoceanic ridges. A schematic example is illustrated in figure 2.14. Note that the differential motion between the plates is in the opposite displacement sense as that anticipated from the apparent offset of the ridge crest. This is a consequence of the fact that the spreading-center segments were formed more or less in their present relative positions (see figure 2.9), rather than having once been a continuous linear feature later displaced by translation along the fracture zone. Note too, that beyond the ridge-ridge portion of the transform, no differential slip exists between the plates. As illustrated schematically in figure 2.14, these inferred dynamics are supported by the pres-

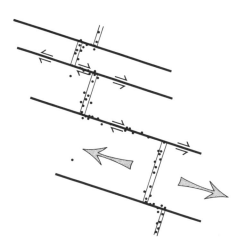

FIGURE 2.14. Diagrammatic representation of an oceanic transform fault, or conservative lithospheric plate boundary. Earthquake epicenters are shown by dots, motions of the plates relative to the spreading center by arrows. Note that transform faults are seismically quiet except for the portions linking adjacent ridge segments, along which active differential slip is taking place.

ence of exclusively shallow earthquake foci only along the active strands of transform faults that connect neighboring segments of spreading centers.

Sedimentary Rocks in the Ocean Basins

Continents contain a thin, nearly continuous veneer of young rocks, but they are made up of more ancient materials as well; the oldest sections of the sialic crust appear to be nearly 3.8 billion years old (see chapter 7). These rocks of great age are preserved imperfectly, and in only fragmentary form, hence the continents bear a rather indistinct, obscure, and incomplete record of early Earth history. In most cases ancient lithologies are confined to the stable continental interiors, or cratons, which are referred to as Precambrian shields.

Prior to the advent of the plate-tectonic concept, geologists wistfully imagined that, in contrast to the fragmentary history of rocks deposited on, or thrust onto the land, the sea floor might contain a complete record of geologic time since the establishment of the hydrosphere (see chapter 7), as embodied in an uninterrupted succession of sediments. If only we could sample them! An implicit assumption, of course, was that the ocean basins represent permanent features of the Earth's crust. We now know that the oceanic crust is formed along midoceanic ridges, and that the sea-floor spreading process carries the crust away to an eventual arrival at a trench where they either offload (are decoupled from the downgoing slab and shuffled up against the margin of the overlying plate) or, much more commonly, disappear into the mantle through descent along an inclined subduction zone. This scenario reflects the flow of the upper portion of mantle convection cells that thereby removes the oceanic crust every 50–200 million years, and the superjacent layers of sedimentary material deposited on it. Therefore, according to plate-tectonic precepts, sediments laid down on such oceanic crust should be exclusively young, compared with the age of the Earth. But are they?

To answer this question, an oceanographic vessel, the Glomar Challenger, capable of operating as a stable drilling platform on the high seas was commissioned in the late 1960s. Results obtained during the decade following abundantly verified the hypothesis of sea-floor spreading. It was demonstrated that only a thin layer of very young sediments occurs in the vicinity of a submarine ridge axis whereas, proceeding in the presumed ocean-floor transport direction, the thickness of the sedimentary cover increases substantially, and the age of the oldest, basal stratum resting directly on the basalt increases commensurately. The most ancient sediments and underlying oceanic crust in the enlarging North Atlantic are Early Jurassic in age (about 190 million years old), reflecting the rifting of Laurasia, as postulated by Wegener. The oldest sediments in the constricting Pacific Basin, situated directly in front of the Mariana trench, are of roughly comparable age. Areal relationships are shown in figure 2.15. The sedimentary contents of the world's marine basins, as well as the underlying oceanic crust, are relatively youthful, geologically speaking. No Paleozoic or older basaltic crust or overlying strata of comparable ages have been found in the world's ocean basins.

Translation distances from the ridge crests, coupled with ages of the oceanic crust deduced from the times of deposition of the immediately overlying sediments, allow spreading rates to be deduced. In general, stable nonsubducting lithospheric plates, such as the American and the Eurasian, are moving away from the midoceanic ridges at one to three centimeters a year; in contrast, plates characterized by descending portions, such as the Pacific, Nazca, and Philippine, are moving six to twelve centimeters per year relative to hot spots and adjacent plates. This velocity relationship suggests that slab "pull" down the subduction zone is a more important factor influencing plate motions than ridge-associated "push."

To summarize, the truly ancient record of the Earth is written solely in the continents, for the continued sea-floor spreading of the ocean

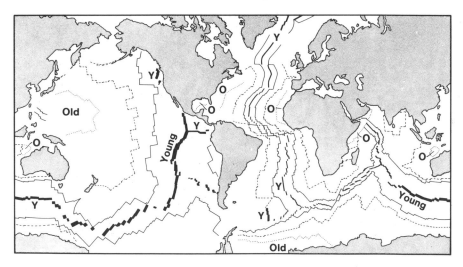

FIGURE 2.15. Age of the oldest sediments (resting directly on oceanic crust) in the present ocean basins (generalized after Pitman, Carson, and Herron 1974).

basins in response to ongoing mantle convection sweeps the sedimentary cover against convergent plate junctions on a cycle of approximately one hundred million years; here the sedimentary debris is at least in part scraped off, or underplated at depth, and both deformed and stored as new portions of an accreting island arc or continental margin.

Wegener Vindicated

Wegener's ideas were rejected in the early part of this century due to a variety of objections, but mainly because the mechanism he advocated to explain continental drift was shown to be physically inadequate to allow the sialic crust to move through the simatic substrate in the manner he erroneously proposed.

Of course, at that time, the geology of the ocean basins was essentially unknown. We now know that both continental and oceanic crusts are embedded in enormous lithospheric slabs, the boundaries of which may—but in general do not—coincide with continent-ocean boundaries. Because the oceanic crust–capped, cooler lithosphere is denser than the underlying, hotter asthenosphere, it is thought that some sort of convective circulation in the mantle is involved in the motions of the plates. We do not know whether rising mantle currents displace the growing, hot, but cooling lithospheric slabs away from the ridges, whether the plates simply slide off the spreading centers (which constitute bathymetric highs relative to the general level of the sea floor), or whether the aging, cold, dense plates are dragged down beneath the trenches. (Probably all three come into play, although the "pull" may be more important than either the gravity-induced slide or "push" because, as previously noted, plates terminating in subduction zones on the average are moving faster than those lacking a downturn.) It is evident that lithospheric sinking does occur, induced by a thermally generated gravitational instability. To preserve constant volume, this requires an asthenospheric return flow. Having said that, we are not much closer to an elucidation of the driving force for the drifting of continents than was Wegener.

What is now universally obvious to earth scientists, however, is that continental drift and sea-floor spreading are taking place today as part of the plate-tectonic process. Incontrovertible evidence includes many new facts, principally obtained from the ocean basins. Supporting observations include bathymetric, magnetic, seismic, and sedimentary data. How plate tectonics may have operated in the distant geologic past will be treated somewhat speculatively in chapter 7.

___ EARTHQUAKES AND VOLCANISM

If you compare a map of historically active and geologically youthful volcanoes as shown in figure 2.16 with a map of seismicity such as figure 2.12, you will be struck immediately by the nearly one-to-one correlation. Earthquakes and volcanoes stand along the margin of the Pacific Basin and form a less distinct, semicontinuous Alpine-Mediterranean-Himalayan-Indonesian seismovolcanic zone as well. Other concentrations occur along the midoceanic ridges, but inasmuch as most of this activity has been submarine, it has heretofore gone largely undetected. One need hardly be a resident of Japan, Chile, Hawaii, or southern Alaska to be aware of the ubiquitous interrelationship: earthquake activity and eruption of molten rock are obviously associated geographically and temporally. Why should this be so?

A class of seismic shocks appears to be a result of frictional resistance accompanying the rise of magma (molten, or partially molten, material) within a conduit, such as the feeder pipe beneath a volcano. This relationship provides the explanation for some vibrational activity preceding eruption in midplate regimes such as Hawaii, and along oceanic spreading centers (e.g., Iceland). However, most earthquakes are a reflection of stress buildup and episodic release in sections of solid rock sliding past one another along a fault, either in the upper mantle or in the crust, or both. Such dynamic physical discontinuities mark all lithospheric plate boundaries, as we have seen. It seems likely that plate boundary faults, especially in regimes characterized by lithospheric extension (i.e., stretching), provide a ready-made "plumbing system" whereby molten material at depth, resulting from partial fusion in the mantle and/or deeply buried crust, can gain access to upper levels of the crust, or the surface. Therefore, the localization of magmatic (igneous) rocks in such linear seismic belts, while not accidental, is not, strictly speaking, a direct result of earthquake energy release.

Most earthquakes originate at shallow depths—say less than fifty kilometers. Plate-tectonic regimes include extensional or divergent boundaries, transform boundaries, and intraplate areas. Within the ocean basins, magmas associated with these midplate and spreading-ridge tectonic settings are dominantly basaltic, but where continental masses have undergone stretching or transform motion, the bimodal association of chemically distinct basalt and rhyolite volcanic assemblages is characteristic. Rhyolites are silicic, alkali-rich volcanic rocks that form in sialic crust in an extensional setting where hot asthenosphere or molten rock rises to shallow depths; many rhyolites are probably derived by

FIGURE 2.16. Modern and recently active volcanoes of the world (simplified after Simkin and Seibert 1984).

partially melting of deeply buried portions of the continental crust in these especially hot environments (see chapter 4).

Shallow-focus, intermediate-focus, and deep-focus earthquakes (hypocenters less than 50 kilometers deep, 50-200 kilometers deep, and 200-700 kilometers deep, respectively) typify convergent plate boundaries. Set back a horizontal distance of about 75–150 kilometers from the trench on the stable, nonsubducted slab, voluminous basaltic and andesitic volcanics and their intrusive equivalents accumulate during the construction of island-arc and continental-margin magmatic provinces. (Andesitic lavas and ash deposits are intermediate in composition between basalt and rhyolite and are essentially chemically equivalent to sial.)

The origin and diversity of rock types with plate-tectonic settings will be treated more fully in chapter 4, and the Circumpacific "Ring of Fire" will be described with the U.S. West Coast as an example in chapter 5. What should be apparent from this brief sketch is that the compositions and areal dispositions of igneous rocks, as well as their association with intense seismic activity, are closely correlated with plate-tectonic regimes.

___ MOUNTAIN BUILDING, PLATE TECTONICS, AND
___ EXOTIC TERRANES

Some mountain chains, such as the Andes, the Sierra Nevada, and all midoceanic spreading centers, are constructed predominantly of surficial lavas, volcanic ash, and molten equivalents that have congealed at depth within the crust. The continents, however, are also typified by orogenic belts that display evidence of great shortening and thickening. These compressional mountains are characterized by pre-existing layered rocks (including, but not exclusively, sediments and volcanics) that have been contorted into a series of folds and fault-bounded blocks of various dimensions.

Wegener related this deformation of previously relatively flat-lying pre-orogenic lithic units to continental drift. He evoked a model of crumpling at the leading edge of the advancing sialic crust. Today we realize that this large-scale wrinkling has been manifested in a way far more complicated than imagined by Wegener. Three different styles of mountain building may be recognized, each of which reflects a particular plate-tectonic regime (convergent, divergent, conservative). All such belts of deformed rocks are correlated with the environs of present or, by inference, ancient plate boundaries. Incidently, we must distinguish topographically high regions that are physiographic mountain ranges, and may or may not contain highly-folded rocks, from structurally contorted true orogenic belts, which may or may not have been eroded to low-lying areas of subdued relief. Deformational structures, rather than height above sea level, constitute the essential hallmark of an orogenic zone. Then, too, the roots of ancient mountain belts may be covered completely or in part by younger, non-orogenic rocks.

Mountain Belts of Convergent Plate Boundaries

By far the most complicated, this type of orogenic belt consists of three distinct structural portions (pictured in figure 2.17): (1) a seaward trench complex and accretionary wedge; (2) a medial forearc basin; and (3) a landward (i.e., continentalward) volcanic/plutonic arc or continental margin. Figure 2.17 presents a schematic but actualistic cross section of such a complex, represented by northern and central California of about 80–120 million years ago. Compare this illustration with the model shown in figure 2.7.

The seaward portion consists of oceanic crust and a prism of sedimentary rocks, individual units of which have been carried beneath the

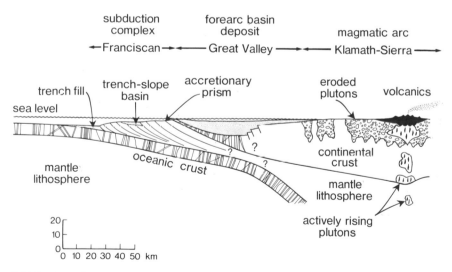

FIGURE 2.17. Diagrammatic cross section through a convergent plate junction illustrating the compound nature of an orogenic belt; the example chosen is northern and central California as it might have appeared near the end of Mesozoic time (after Dickinson et al. 1982).

overriding, stable, nonsubducted lithospheric plate. Decoupling of this largely sedimentary debris in the vicinity of the oceanic trench, or deep, and on and beneath its landward side, results in the accumulation of an accretionary wedge, or subduction complex. Low-angle faults in which the overlying parts move upward relative to the underlying segments are termed thrust faults; a series of subparallel thrust faults roughly parallels the boundary of the descending slab and causes imbrication (overlapping and shuffling of the section at the margins) and folding within the accretionary wedge. As this section thickens and buoyantly rises, low-angle faulting of opposite sense takes place.

Immediately landward from the trench offscrappings, and above the underplated region deep beneath the overlying wedge, lies a column of forearc basin sediments in the so-called arc-trench gap. These sediments, like those of the seaward trench deposit, have been shed off predominantly from the growing volcanogenic arc. Resting on the nonsubducted lithospheric plate, arc-trench gap strata in general are much less deformed than the contorted melange complex residing in the jaws of the subduction zone.

The landward arc is yet more complicated, because such belts typically contain remnants of the old presubduction continental margin, plus a host of magmatic rocks coeval (contemporaneous) with arc formation. The melts rise from the mantle above the downgoing slab, and transport heat into the crust, thereby causing partial liquification of the most fusible, deeply buried portions of the sial. Some of the buoyant molten material rises but solidifies at depth, whereas other portions gain access to the surface and vent as explosive, ash-laden clouds or as less violently extruded lavas.

Fracturing, faulting, and deep-level ductile deformation of all three domains that constitute convergent plate boundary orogenic belts are common. Because substantial motions take place at an angle to the plane of cross section illustrated in figure 2.17, lateral shuffling of the various lithologic regimes also occurs. Moreover, the sea-floor-spreading process modifies the accretionary margin by bringing exotic oceanic crust and fragments of far-traveled, unrelated sialic crust to the convergent junction, and removing native segments, thus providing additional complications to this compressional orogenic system. Rearrangement due to this transform component of movement results in the construction of a collage of terranes—individual fault-bounded blocks—some formed more or less in place, others truly exotic (far-traveled). Thus, the evolution of a continental margin may take place through the addition of far-traveled, unrelated terranes as well as through the generation of new sialic crust within the convergent plate margin regime.

Mountain Belts of Extensional Plate Boundaries

A narrow central rift zone developed in the oceanic crust characterizes midoceanic spreading ridges (see, e.g., figure 2.6), reflecting the 180-
degree divergence of back-to-back lithospheric slabs. Where extensional
zones occur within the much thicker continental crust and subjacent
mantle lithosphere, the attenuation typically is distributed over a much
broader region. As shown in figure 2.18, faulting involves jostling, lateral
stretching, and downward relative motion of the crustal blocks, resulting in block-fault mountains, tilting, and thinning of the crust. Such
faults are known as normal faults. Examples of continental extensional
zones include the Rhine Graben (graben is German for "trench" or
"grave") of West Germany, the Basin and Range Province of Nevada and
western Utah, and the East African and Dead Sea Rift systems. Thermal
expansion due to close approach of asthenosphere to the surface causes
uplift early on; ultimately, the rifted area may subside to form a new
ocean basin, such as the Red Sea. Rise of a column of asthenosphere,
bifurcation near the surface, and horizontal flow all seem to be indicated
in extensional plate-tectotic settings. Uplift followed by such crustal
attenuation appears to presage the incipient stretching and breakup of
continents, as may now be taking place in the Rio Grande Rift of New
Mexico, and the East African Rift system.

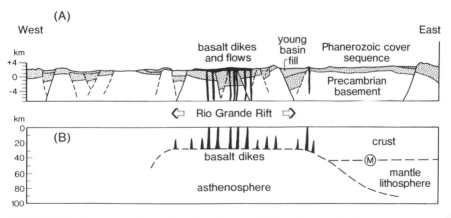

FIGURE 2.18. Cross section of the Rio Grande Rift, north-central New Mexico, an
example of an incipient divergent plate junction—or, at least, stretched, thinned
lithosphere—manifested in continental crust (simplified after Seager and Morgan 1979). The upper crust is illustrated with vertical exaggeration in (A); the crust
and upper mantle are shown at the same vertical and horizontal scales in (B).

Mountain Belts of Conservative Plate Boundaries

Ideally, transform-type plate junctions involve neither convergent nor divergent motion. Thus disposed, they would have little effect on the transected oceanic or sialic crust other than to divide it into two portions, juxtaposing dissimilar rock units across the slip zone. However, transform plate boundaries that transect old, geologically heterogeneous and complex continental crust tend to involve deflections contrasting markedly with the straight planar features of the oceanic fracture systems.

As illustrated schematically in figure 2.19, these bends in complex sialic crust are of two types: mass excess and mass deficient. During movement along a curved fault, the former results from a constraining bend, in which the impingement of excess material is accommodated by compressive deformation, localized mountain building, and thickening of the continental crust; the latter generates a topographic depression or pull-apart basin at a releasing bend during slip along the transform. Thus continental transform boundaries (strike-slip faults) are characterized by straight, mass-conservative horizontal-slip segments, and by locally deformed orogenic zones showing adjacent upthrust blocks (compressional folds and high-angle thrust faults) and downdropped blocks (sag basins and normal faults). Where fractures and faults extend to great depth in the continental crust, and into the upper mantle, molten material may gain access to the transform zone as igneous intrusions and extrusions. These result in what have been called "leaky transforms."

Exotic Terranes

We have seen that the differential motions of the lithospheric plates results in the dispersal of some formerly contiguous crustal units and in

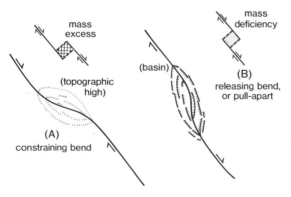

FIGURE 2.19. Constraining (A) and releasing (B) bend regions along continental transform faults. Insets diagrammatically show regions of mass excess and mass deficiency, exaggerating angular relationships for clarity.

the juxtaposition of other far-traveled geologic entities of contrasting crustal origin and history. The rifting apart of pre-existing continents and the rearrangement and suturing of microcontinental fragments, island arcs, and oceanic crust after an episode of sea-floor spreading and drift results in the construction of new sialic collages. Such assemblies consist of unrelated, fault-bounded terranes, some of which actually may be oceanic in origin.

The principles of plate tectonics, as illustrated in figures 2.6, 2.7, 2.9, and 2.19 require that plate junctions mark the dynamic boundaries of individual crust-capped lithospheric slabs, and that all plates move differentially with respect to one other. In a real sense, the continents are all exotic, unrelated terranes, having been transported differentially relative to one another (figures 2.1 through 2.4). On a smaller scale, the area west of the San Andreas fault is far-traveled relative to the rest of California (figure 5.5). So is India relative to the rest of Eurasia. Some earth scientists regard much of the western cordillera of North America, and the Indonesian archipelago as constituting amalgamated assemblages of unrelated terranes (see figure 5.2). In any event, we must regard the continents as complexes consisting of both native and foreign crustal units.

MANTLE CIRCULATION AND THE DRIVING FORCE OF PLATES

Mountain building, earthquakes, volcanic activity, crustal growth, seafloor spreading, continental drift, and subduction are all surficial manifestations of flow within the somewhat deeper Earth. This circulation, or convection, represents the mechanism that Wegener sought, but never discovered, in his attempt to prove the reality of continental drift. Even today, mantle flow, thought to be on the order of a few centimeters per year, has not been measured directly.

Convection results from the conditions in a relatively low-viscosity medium whereby a dense upper layer surmounts a less dense substrate; on perturbation, overturn occurs as a consequence of the gravitative instability. This density inversion and resultant bodily flow reflect the fact that as a solid (with or without interstitial melt) cools, it contracts; accordingly, lithosphere, the outer, cooler rind of the Earth, has a higher specific gravity than the hotter asthenosphere just below it. With increasing depth in the Earth—for the same bulk composition, and in the absence of phase changes—the progressively hotter mantle may become sufficiently less dense and sufficiently less viscous to be buoyant and capable of overturn. Below the transition zone, dense minerals present in the lower mantle (which may also be richer in iron than the upper mantle) evidently are gravitively stable.

Therefore, we can recognize two kinds of motive force for upper-mantle convection: (1) near-surface descent of lithospheric plates, occasioned by their slipping laterally off bathymetric elevations such as midoceanic ridges, as well as sinking in the vicinity of trenches; and (2), upward-welling, buoyant mantle flow due to deeply buried thermal anomalies. The abundantly documented, ubiquitous occurrences of oceanic ridges, subduction zones, and mantle plumes (hot spots) demonstrate that both types of process are operating today, and probably also attended past plate motions. The relative magnitudes of the near-surface and deep-driving forces that contribute to the upper mantle circulation cells are still a matter of debate, however.

☰ 3
☰ MINERALS: BUILDING BLOCKS OF ☰ ROCKS

Earth scientists study the origin and evolution of the outermost portions of our planet through investigation of the rock record. The long and involved history of the Earth is recorded and imperfectly preserved chiefly in the lithosphere—very much less so in the hydrosphere, atmosphere, and deep interior of the Earth. Regarding the lithosphere, the crust is, of course, the most readily and completely accessible part. Because rocks are made up largely of minerals, geologists, geophysicists, and geochemists must develop an appreciation for the nature of these constituents in order to understand the diverse origins of rocks. Of obvious practical value are terrestrial mineral resources (see chapter 6), the largely inorganic materials, as well as fossil fuels, necessary for the continued support and well-being of civilization, and, in some cases, a prerequisite for life. Just what are these earth materials—minerals, mineraloids, crystals, and rocks?

☰ MINERALS: WHAT ARE THEY AND WHY STUDY THEM?

A mineral is a naturally occurring, inorganically produced solid possessing a characteristic chemical composition or limited range of compositions, and a systematic, three-dimensional atomic order (crystal structure). It is invariant in its chemical and physical properties, or exhibits a restricted range in characteristics. A mineral is a homogeneous phase; that is, it is not separable by mechanical means into two or more substances (two or more phases) of contrasting chemical and/or physical properties. The properties in question may be, for example, specific gravity, solubility, hardness, color, magnetic susceptibility, melting temperature, electrical or thermal conductivity. Minerals are solids, in contrast to liquid or gas phases. They are compounds of invariant composition, such as quartz, SiO_2; or, they have compositions that range between fixed values, such as the olivines, $(Mg,Fe)_2SiO_4$. The latter exhibits compositional variation from a pure magnesium silicate, Mg_2SiO_4, to a pure iron silicate, Fe_2SiO_4. Minerals are constructed of atoms, systematically located and exactly repeated in three dimensions. The atomic configuration is termed the crystal structure; for most minerals, the arrangement of atoms and the periodicity (repeat distance) are different in different directions. All substances having a regular, ordered atomic structure are said to be crystalline; therefore, all minerals are crystalline.

A mineraloid is any naturally occurring solid or liquid that lacks a systematic periodic arrangement of the constituent atoms. Mineraloids are therefore noncrystalline, or amorphous; that is, they do not possess a crystal structure. Volcanic glass, amber, coal, and petroleum are examples of mineraloids. In the broad sense they belong to the mineral kingdom, which accounts for the usage of the term "mineral resource" to include mineraloids. Note that some mineraloids are biologically produced. On an atomic scale, mineraloids possess some chemical polymerization, or short-range ordering, but a rigorous, long-range, three-dimensional atomic structure is lacking.

A crystal is any grain of crystalline material bounded by crystal faces; the latter are planar surfaces bearing a definite geometric relationship to the atomic arrangement. For example, consider the beautiful planar terminations bounding "rock crystal" quartz, SiO_2, or "dogtooth spar," a variety of calcite (figure 3.1, A and B respectively). These crystal faces reflect the manner of crystal growth, and are disposed with specific angular relationships to the atomic structure of the mineral (see pp. 83–89). It would be possible to grind planar facets at any orientation to the regular, internal, ordered atomic arrangement of such crystals, as is done to produce "cut" diamonds, but, in the general case, these would not be crystal faces and would not be viable growth surfaces. Mineral grains bounded by crystal faces are termed euhedral, whereas irregularly terminated grains are called anhedral.

A rock is a naturally occurring, cohesive, multigranular aggregate of one or more minerals and/or mineraloids. Units considered generally must be large enough to constitute an important part of the solid Earth. Obviously, gradations exist from a mountain range composed of, for example, granite, to granitic seams and veins of infinitesimal thickness cutting through another rock. Actually, what is important is the fact that the constituent grains either of the same or of different substances, owe their present association to a common genetic process, the origin of the rock itself. The genesis and diversity of rocks will be described in the next chapter; only a few definitions are presented here.

Broadly speaking, geologists recognize three main rock-forming processes, and therefore three principal classes of rocks. In the first, molten rock-forming material, or magma, solidifies either to glass or to an aggregate of one or more types of minerals, or to a combination of glass and minerals: such rocks are termed *igneous*. Ignis is the Latin word for "fire"; igneous rocks are, crudely put, formed from fire. Lava flows and ash falls are obvious examples of igneous rocks. *Sedimentary* rocks, products of the second process, consist of mechanically or organically accumulated fragments of pre-existing rocks and minerals, as well as chemical or biochemical precipitations from a fluid medium. As indicated by the root sedimentum, Latin for "a settling," sedimentary rocks have had their constituent particles settle out, generally either subaerially or, more generally, subaqueously. Some examples of sedimentary deposits are coquina (shell beds), and stream gravels; when these become coherent, cohesive rocks, they are called shelly limestone and conglomerate, respectively. The final class is the *metamorphic*, which includes all those rocks whose original minerals or textures, or both, have been altered markedly by recrystallization and/or deformation subsequent to the original formation of the precursor rock. Metamorphism may take place at considerable depth within the Earth or, in some cases, fairly near the surface due to adjacent emplacement of a hot igneous mass. The Greek word meta is translated as "successive," "after," or "change"; hence a metamorphic rock represents a later configuration of minerals and/or

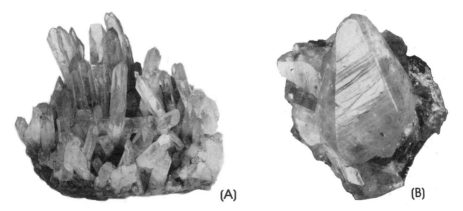

(A) (B)

FIGURE 3.1. (A) Quartz, variety "rock crystal," UCLA collection; (B) calcite crystals, so-called "dog tooth spar," UCLA collection (Ernst 1969).

textures different from those of the original rock. Greenstone, marble, and slate are familiar examples of metamorphic rocks, derived from basalt, limestone, and mudstone, respectively.

☰ PRINCIPLES GOVERNING THE ARCHITECTURE OF MINERALS ☰ AND CRYSTALS

The three-dimensional arrangement of atoms—in other words, the crystal structure—determines the chemical and physical characteristics displayed by individual minerals. Of course, inasmuch as all minerals of a particular species possess a common crystal structure, all will exhibit the same physical and chemical properties. It is therefore illuminating to consider factors that influence the type of crystal structure developed for a given composition. But before we can understand the atomic linkages in a crystalline material, we must review the nature of atomic particles and the forces that bind atoms together.

Structure of the Atom

The Danish physicist Niels Bohr proposed the atomic model described here (appropriately termed the Bohr atom). The atom is the smallest divisible unit retaining the characteristics of a specific element. It consists of an atomic nucleus and surrounding electrons confined to specific orbitals, or energy levels; in figure 3.2 the electron orbitals are shown schematically as a series of concentric spherical electron shells. The nucleus has a diameter of approximately 10^{-13} centimeters; in contrast, the diameter of the enveloping electron shells is about 10^{-8} centimeters, or 100,000 times the diameter of the nucleus. In some respects, then, the structure of the atom is crudely similar to that of a tiny solar system, with the nucleus as a central star, and the electrons as planets.

The nucleus contains two principal kinds of atomic particles: protons, each with a positive charge, and uncharged neutrons. Both types of particle have approximately the same mass, a unit value of one. The number of protons, Z, defines the atomic number of an atom; each chemical element is distinguished by a different value of Z. The sum of protons and neutrons determines the characteristic mass, or atomic weight, of an element. Individual atoms of the same element (same number of protons, hence same Z) having different numbers of neutrons are termed isotopes of that particular element. For instance, oxygen ($Z = 8$) has three isotopes. The nucleus of the most common species contains eight protons and eight neutrons and is referred to as ^{16}O; one of the two rarer isotopes, ^{18}O, carries eight protons and ten neutrons in its nucleus. There are also three isotopes of hydrogen (Z = 1): common hydrogen, ^1H; deutrium, ^2H; and the very rare tritium, ^3H. The chemical behavior of an atomic species is a function of the number of protons, hence nuclear charge, rather than its complement of neutrons, thus mass. For this reason, chemically distinct elements are indicated by their Z numbers, not their atomic weights.

A negatively charged electron cloud encircles the nucleus. Each electron has a charge equal to that of a proton but of opposite sign; the mass of an electron, however, is only approximately 1/1,837 that of a proton. Therefore, its mass is negligible and, to a first approximation, electrons can be ignored when calculating the atomic weight of an element. On a statistical basis, electrons are restricted to specific energy levels or orbital shells, roughly concentric to, or at least symmetrically disposed about, the nucleus. These levels differ by discrete amounts of energy, or quanta. The general principle that atomic particles can exist only with certain energy configurations was postulated by the German physicist Max Planck and represents the cornerstone of quantum theory. According to this theory, energy exists as discrete bundles on the atomic scale, not as an infinitely divisible energy spectrum. Thus, electrons surrounding the nucleus can occuppy only specific energy levels—i.e., electron shells—differing from one another by a discrete number of quanta. Although it is important to remember that, physically, electrons exhibit wavelike as well as particulate aspects, we will consider electrons simply as negatively charged particles.

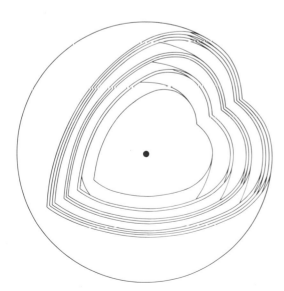

FIGURE 3.2. Schematic view of the Bohr atom, showing the atomic nucleus and the enveloping electron cloud (Ernst 1969). Although the electrons are shown here as confined to specific spherical shells, in fact the orbitals, although symmetrical, are much different in their disposition (see figure 3.4). Nucleus not to scale.

When there are as many electrons in the surrounding electron cloud as protons in the nucleus, the net charge is zero, and overall the atom is electrically neutral. Of course, higher atomic number atoms possess more numerous, high populated electron orbitals compared with atoms characterized by a low Z. The innermost or K (the first) shell, can contain a maximum of two electrons. Outer orbitals, which represent higher energy levels, are indicated as L (second), M (third), N (fourth), and so on. Eight electrons can be accommodated in the L shell, eighteen in M, and thirty-two in N; usually, the higher shells are incompletely filled. As another complicating factor, subshells of contrasting energy occur within a specific shell. Listed in order of increasing energy, they are subshells s, p, d, and f. In some cases, a suborbital of a higher shell number—for instance, 4s of the N shell—will be filled before electrons enter 3d of the M shell, because an electron occupying the 3d subshell actually has a higher energy than one occupying 4s.

As shown in figure 3.3, a further complication arises because orbital energy depends on the total nuclear charge, Z. The electron capacities of the subshells are as follows: $s = 2$; $p = 6$; $d = 10$; and $f = 14$. Although these shells and subshells are shown as spherical in the simple Bohr atom

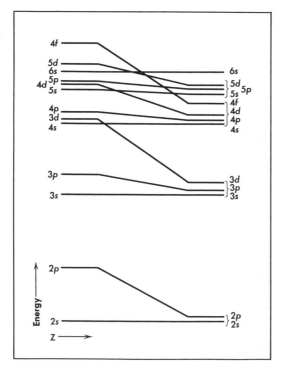

FIGURE 3.3. Energy levels of electron subshells as a function of increasing atomic number (Ernst 1969).

(figure 3.2), the geometries of the electron cloud for specific suborbitals, while highly symmetrical, actually depart markedly from this model; some typical orbital relations are illustrated in figure 3.4.

In the ground state of the atom, electrons occupy the lowest energy configuration possible. For a neutral atom in an excited, or higher-energy state, one (or more) of the inner-orbital electrons may be missing; instead, an extra electron (or more) will occupy an outer, incompletely filled shell.

Another sort of higher-energy state arises when an initially neutral atom either loses or gains outer orbital, or valence, electrons. It thereby acquires a positive or negative charge, and in this condition is termed an ion. Positively charged ions, which have lost electrons, are called

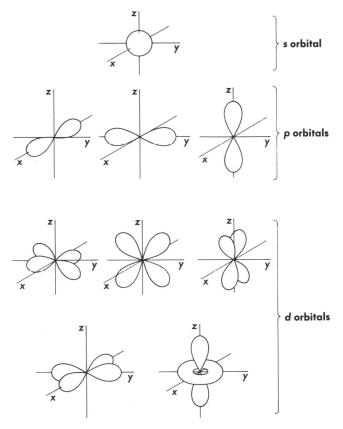

FIGURE 3.4. Characteristic, diagrammatic subshell electron orbitals (Ernst 1969). The *f* orbitals are partially filled only in the higher atomic number atoms, are more complicated, and are therefore not illustrated.

cations because they are attracted toward a cathode, the negatively charged terminal of an electrical cell; anions carry excess negative charges and are attracted by a positively charged plate, the anode. Electronegativity measures the ability of an atom to attract electrons to itself. Those atoms characterized by low electronegativities, such as the alkali metals lithium, sodium, potassium, rubidium, and cesium, readily form cations because the valence electron is easily relinquished. In contrast, the halogens flourine, chlorine, bromine, and iodine have high electronegativities and readily form anions because they are avid electron acceptors.

Bonding

So far we have confined our attention to individual atoms. Yet, with the exception of the noble gases—helium, neon, argon, krypton, xenon, and radon—natural materials consist of atoms that are bonded to one another to form molecules. The inert noble gases, which are otherwise of minor importance in mineralogy and petrology (the studies of minerals and rocks, respectively), suggest an explanation for the mutual attraction that occurs among both like and unlike atoms. Atomic structures of the noble gases are stable, minimal-energy configurations because the outer, or valence, orbitals are already completely filled. Therefore, these elements generally occur as single (unbonded) atoms.

In contrast, the bonding together of other elements, which as neutral atoms have incompletely filled outer orbitals, represents an attempt to approximate the noble gas condition of electron-saturated orbital shells. This is why atomic sodium, characterized by a single 3s electron, readily loses this electron (becoming positively charged), and chlorine which, as a neutral atom, possesses five 3p electrons, gains one (becoming negatively charged): as ions, both positively charged sodium (Na^+) and negatively charged chlorine (Cl^-) exhibit noble gas–type outer electron orbital configurations, with totals of eight and eighteen electrons respectively (see figure 3.6). The ions are mutually attracted because of opposite charges; the nucleus of one attracts the electrons surrounding the other and vice versa. These attractive forces are inversely proportional to the square of the distance separating the nuclei but are modified somewhat by the screening effect of the negatively charged electron cloud encircling the positively charged nucleus. Atoms are prohibited from extremely close approach to one another partly because, as the interatomic separation decreases, the electron clouds are mutually deformed; but principally because the strongly concentrated positive charges in the approaching nuclei generate intense repulsive forces. Such relations are shown in figure 3.5. The equilibrium distance between nuclei represents the minimum energy configuration and the stable bond configuration.

What is the nature of these forces binding the atoms together? Depending on the degree to which nuclei attract or give up valence-shell electrons, we can distinguish four basic bond types: ionic; covalent; metallic; and van der Waals. For simplicity, most mineral structures are treated as though they were purely ionic, but actual bonding is generally intermediate or mixed in character among the several bond types.

Ionic bonding results from the electrostatic attraction of oppositely charged ions, as already introduced above. Cations attain a noble-gas configuration through electron loss, anions by electron gain. As examples, the electronic structures of Na^+ and Cl^- are shown in figure 3.6. For most purposes, ions may be imagined as spheres of nearly fixed radii; however, in the presence of a strong electronic field, they are deformed, or polarized, to a certain extent. In crystalline NaCl, which occurs as the mineral halite, each sodium ion is surrounded by six

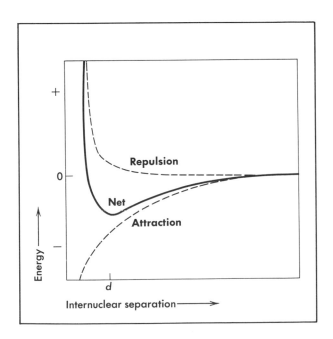

FIGURE 3.5. Schematic repulsion curve indicates force between neighboring nuclei; attraction curve shows force between nucleus and electron cloud of neighboring atom; d gives the minimum energy configuration, or equilibrium internuclear separation (Ernst 1969).

chlorines, and contrariwise (see figure 3.15). The ions, considered as incompressible spheres, are packed together in such a way as to produce nearly minimal void space. The structure is coherent, not because there are discrete bonds linking pairs of ions together, but because each ion is attracted by its six nearest neighbors, all of which are of opposite charge.

In the periodic table of the elements, printed at the back of the book, vertical columns of atoms denote atomic groups; within one group, all neutral elements have the same number of electrons in their valence orbitals, but different numbers of shells. (The horizontal rows are termed atomic series; within any one atomic series, all elements have the same number of electron-energy levels, but different extents of filling of the valence orbitals.) Elements of atomic groups I and VII readily lose or gain an electron, respectively, to complete the noble-gas structure. Group II and group VI elements lose or gain two electrons, respectively, but inasmuch as this is more difficult than losing or gaining a single electron, they do not exhibit strictly ionic character. Because elements of atomic groups III, IV, and V tend to retain their electrons more tenaciously, covalent bonding, now to be described, becomes important for substances containing these elements.

The completion of valence orbitals by electron sharing (rather than electron donating or accepting) characterizes *covalent bonding*. In this type of attraction, electron clouds interpenetrate. The configurations of two gas species, molecular oxygen and methane, are shown schemati-

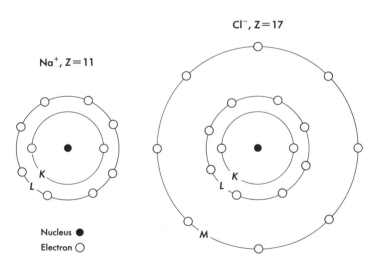

FIGURE 3.6. Diagrammatic representation of electron-shell population for Na^+ and Cl^- ions (Ernst 1969). Atomic particles are not to scale. The bond attracting these contrasting chemical species is ionic in nature.

cally in figure 3.7. These species, which contain two and five atoms, respectively, are called molecules. The examples are not solids, but bonding in many minerals also is partially or even chiefly covalent in character. The number of coordinating nearest neighbors is restricted not by size and charge considerations as in the case of ions, but by the small number of discrete bonds (shared electrons) that are necessary to complete the inert gas–type structure.

Elements of the periodic table of the elements that display a covalent nature occur especially but not exclusively in atomic groups III, IV, and V. In any one series, as atomic number increases, the radius of the neutral atom decreases. In effect, the increased charge on the nucleus binds the enveloping cloud of electrons more strongly to itself. Although group I elements have low electronegativities, and accordingly give up their sole valence electron easily, groups III, IV, and V are typified by higher values, and thus attract outer-orbital electrons much more effec-

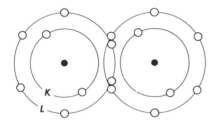

(A) O_2, $Z=8$

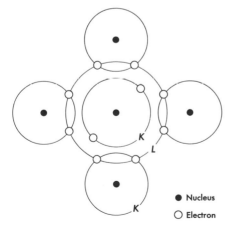

(B) CH_4, $Z=6, 1$

● Nucleus
○ Electron

FIGURE 3.7. Diagrammatic representation of electron-shell populations for (A) O_2 and (B) CH_4 (Ernst 1969). Interpenetrating electron orbitals reflect electron sharing. Atomic particles are not to scale. The bonding in both O_2 and CH_4 is covalent in nature.

tively. The American chemist Linus Pauling recognized that the difference in electronegativity between any two mutually attracted elements provides a measure of the bond character. As shown in figure 3.8, the greater this difference in electronegativity, the more ionic the bond.

Metals are closely packed assemblages of positive ions permeated by a negatively charged continuum of electrons. Attractive forces between the free electrons and the metal ions provide structural cohesion. *Metallic bonds,* like ionic bonds, are not localized geometrically, in contrast to the discrete bonding directions of covalent atoms. The great mobility of electrons in metals accounts for high thermal and electrical conductivities of these substances; softness and malleability also appear to be related to lubrication of the atoms by the freely moving swarm of electrons, or "electron gas." In contrast to ionic bonds, metallic bonds bind together identical cations (metallic elements), or cations closely similar in size and charge (metallic alloys). Size is thus a critical constraint in metallic structures, and the arrangement of anions coordinated around cations need not yield local charge balance, as with ionic structures.

Van der Waals bonds (also known as residual bonds) are exceedingly weak attractive forces that arise from nonuniform charge distribution (electronic asymmetry) or the polarization of otherwise neutral atoms, molecules, or ionic complexes. Such residual bonds, for example, give liquid or solid argon its weak cohesion at extremely low temperatures, in the absence of free, shared, or donated electrons.

Bond types directly influence the physical properties of materials.

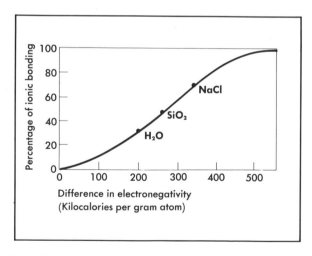

FIGURE 3.8. Bond type as a function of differences in electronegativity, (Ernst 1969).

Where bonds are weak, substances usually are soft and have low melting temperatures, or they decompose to form other minerals, because the binding forces are easily disrupted. The stronger the bond, the harder and more refractory the substance typically becomes. Hardness is strictly a function of solid-state bond strengths, but the temperature of melting is also related to cohesive forces in the liquid as well as in the mineral; nevertheless, a low melting point generally characterizes weakly bonded solids. Many mineralogic structures consist of various elements displaying contrasting bond types; in general, the weakest bonds determine the principal physical properties of such materials.

Ionic Radii

Within any one group (vertical column) of the periodic table of the elements, the radii of the ions increase as atomic number increases, and within any one series (horizontal row), the radii of the cations decrease as Z increases, just as for neutral atoms. In situations where several valence states are possible, such as Fe^{+3} and Fe^{+2}, radius decreases with increasing positive charge. As you may suppose, this is a consequence of the fact that, for the higher positive valence, the fixed numbers of nuclear protons have an attractive force spread amongst fewer orbiting electrons. Finally, the arrangement and number of surrounding anions (see below) slightly modifies the effective size of the cation. An Al^{+3} surrounded by four negatively charged ions (that is, the cation is central to a tetrahedral array of anions) has a slightly smaller effective radius than Al^{+3} surrounded by six anions (an octahedral anion array), because tetrahedral coordination provides a smaller central cavity than does octahedral coordination: accordingly, the electron clouds surrounding the ions must be somewhat deformed, or polarized, to attain the tetrahedral configuration. Ionic radii for some of the more abundant cations and anions are listed in table 3.1 (see also figure 3.9).

For certain elements in silicate structures, large discrepancies exist between conventionally recognized values and the observed interatomic separations. For instance, silicon-oxygen and aluminum-oxygen distances in many rock-forming minerals are considerably less than addition of the appropriate ionic radii would suggest; on the other hand, carbon-carbon bond distances in diamond and graphite are greater than would be predicted from the data in table 3.1. These discrepancies reinforce the point that mineral structures are not wholly ionic in their nature.

It should be noted that the important anions of the Earth's crust and upper mantle, O^{-2}, OH^-, F^-, Cl^-, Cl^-, and S^{-2}, are much larger than the common cations (with the sole exception of the monovalent, posi-

TABLE 3.1. Radii of the Common Ions

Ion and Charge	Radius in Å	Number of Anions Surrounding Cation
OH^-	1.40	—
C^{+4}	0.15	3
O^{-2}	1.35	—
F^-	1.29	—
Na^+	1.02	6
Na^+	1.18	8
Mg^{2+}	0.72	6
Al^{+3}	0.39	4
Al^{+3}	0.54	6
Si^{+4}	0.26	4
S^{-2}	1.84	—
S^{+4}	0.37	6
Cl^-	1.81	—
K^+	1.38	6
K^+	1.59	10
K^+	1.64	12
Ca^{+2}	1.00	6
Ca^{+2}	1.12	8
Ti^{+3}	0.67	6
Ti^{+4}	0.61	6
Mn^{+2}	0.83	6
Mn^{+3}	0.65	6
Mn^{+4}	0.53	6
Fe^{+2}	0.78	6
Fe^{+3}	0.65	6
Zn^{+2}	0.74	6
Sr^{+2}	1.18	6

SOURCE: After Shannon and Prewitt 1969, and Shannon 1976.

tively charged ion, K^+). This being the case, we may visualize mineral structures as relatively closely packed arrangements of anions, with enough cations in the interstices to maintain local charge balance. Volumetrically, anions dominate the structures of minerals outside the Earth's core.

Pauling's Rules

Let us now consider the ordered three-dimensional arrangement of atoms, ions, and molecules in crystalline materials. For simplification, we will consider minerals as strictly ionic structures, even though we know that bonding is only partly ionic. Five observations, known collectively as Pauling's Rules, apply to such structures, although they are more or less applicable to other types of bonded structures as well, excepting metals and organic complexes. They are as follows:

1. A group of anions, known as a coordination polyhedron, is formed about each cation. The outlining edges of the polyhedron may be visualized as lines connecting pairs of points, each point representing the center of an anion, as shown in figure 3.9. Three surrounding anions define a triangle, four a tetrahedron, six an octahedron, eight a cube, and so on. In some cases the polyhedron is highly symmetrical, or regular, while in others it is of lower symmetry. The cation-anion distance is determined by the sum of the respective radii, the ions being considered

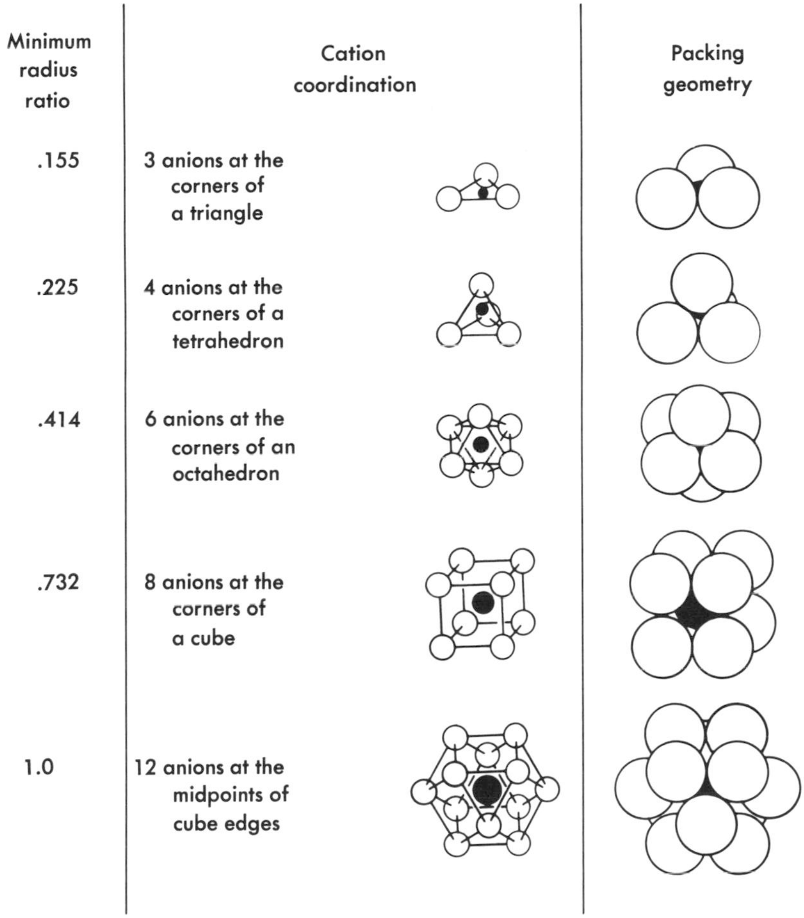

Minimum radius ratio	Cation coordination		Packing geometry
.155	3 anions at the corners of a triangle		
.225	4 anions at the corners of a tetrahedron		
.414	6 anions at the corners of an octahedron		
.732	8 anions at the corners of a cube		
1.0	12 anions at the midpoints of cube edges		

FIGURE 3.9. The interrelationship between cation packing geometry and radius ratio (after Dennen 1960, table 2.4).

as essentially rigid spheres. The cation coordination number—that is, the number of nearest-neighbor anions—is a reflection of the ratio of cation radius to anion radius, or radius ratio. If the central cation is small relative to the surrounding negative ions, only a few of these anions will be able to cluster around the central cation, as is clear from the figure.

2. For a stable structure, the total strength of valence bonds reaching a single anion in a coordination polyhedron from all neighboring cations is equal to the total charge of that anion. Consider the structure of halite (figure 3.15). Each sodium has been stripped of its single valence electron, and is surrounded by six negatively charged chlorine ions; thus each of these Cl^- ions receives, statistically, one-sixth of an electron from every adjacent sodium. Each chlorine also has six nearest-neighbor sodiums, so the total charge of the anion, -1, is satisfied by the electron contributions from all six of the surrounding cations. Of course, the electrons have not been divided up into sixths; the point is that ionization results in completed electron shells and in the generation of moderately strong, nondirectional attractive forces between sodium and chlorines.

3. The polyhedra in a structure tend not to share edges (pairs of anions), and especially not faces (three or more anions), because sharing decreases the distance between the mutually repulsive central cations. If edges are shared, the shared edges are shortened; this results in an increase in the cation-cation separation at approximately right angles to the shared edge.

4. Because such sharing decreases the stability of a structure, cations with high valences and small coordination numbers rarely share polyhedral edges and faces with each other, for if they did, these mutually repulsive cations would be brought into close proximity.

5. The number of structurally different kinds of atoms in a specific structure tends to be small. Another way of stating this rule is that, for mineral structures, characteristically, there are only a few types of contrasting cation and anion sites. Depending on the dimensions and on the local electrostatic configuration of each site, any one of several different chemical species may be accommodated, but the nature of the structural position remains virtually the same.

It is evident that the spatial arrangement of the atoms in a structure reflects local charge and size requirements. Many mineral structures can be explained on the basis of these rules, but it is very difficult to predict the structure of a mineral given only the sizes and charges of the constituent ions, because the spatial-energetic relations are quite complicated and as yet are only imperfectly understood.

Coordination Number. We have seen that the ratio of cation radius to anion radius determines the number of nearest anion neighbors surrounding a central cation. Figure 3.9 presents minimum radius ratios for specific coordination numbers and geometries of the polyhedra. For example, figure 3.10 illustrates why a cation with a radius about 40 percent as large as that of the adjacent anions will have six nearest neighbors; the geometric calculations demonstrating that the central cation has an ionic radius 41.4 percent that of the surrounding anions (assuming the packing of incompressible, spherical ions) is given in the figure. For cations and anions of nearly equal size, the coordination number is 12. As an analogy, consider the initial triangular arrangement of fifteen billiard balls prior to the start of a game of pool. Any ball—say, the eight ball—near the center of the triangle, has six neighbors in one plane; to complete a three-dimensional coordination of 12, three additional balls would have to be placed above it in a similar closest-packing configuration, and three more underneath, yielding a total of twelve nearest neighbors surrounding the eight ball. This example illustrates the point that, like the eight ball, large cations that have ionic radii comparable to that of surrounding anions (e.g., potassium), have coordination numbers approaching 12; in closely packed metals, the ions, being similar or identical in size, normally have this coordination number too.

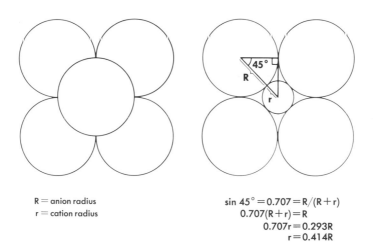

R = anion radius
r = cation radius

$\sin 45° = 0.707 = R/(R+r)$
$0.707(R+r) = R$
$0.707r = 0.293R$
$r = 0.414R$

FIGURE 3.10. Packing of incompressible spheres illustrating a cation/anion radius ratio of 0.414 for octahedral (sixfold) coordination of the central cation, (Ernst 1969).

In general, the radius ratio calculated for a particular coordination arrangement indicates the minimum size of cation that can be accommodated. Ions tightly packed together have low energy; in contrast, high energy characterizes a cation "rattling about" in too large a structural position because of an excessive separation of opposite electrical charges. Thus the peripheral anion group tends to collapse around a small, central, positively charged ion. As progressively bigger cations occupy a specific site, the polyhedron of surrounding anions becomes deformed and inflated so that the central cation can be accommodated.

Polymorphism. This is the situation in which different packing arrangements of constituent atoms (i.e., different structures) occur for a specific composition. Two or more minerals of the same chemical composition but with contrasting structures are thus polymorphs—which is to say that several forms occur. Perhaps the most familiar examples are the polymorphs of carbon. As you might expect, the carbon atoms are densely packed together in diamond, whereas graphite has a more open structure. The contrasts in physical properties are quite marked, as everybody knows.

Solid Solution. Minerals of different chemistries possessing the same structures and closely similar sizes of the corresponding ions generally show a range of compositions intermediate to the pure end members. This range, known as solid solution, is analogous to liquid solution except for the more systematic, three-dimensional atomic periodicity in solids. In some minerals, the solid solution may have any composition between—or among—those of the end members, but in other cases the compositional range of the mineral is limited.

Three principal types of solid solution may be distinguished. The type known as substitution solid solution is illustrated by hypersthene, where six-fold coordinated magnesium and ferrous iron are almost completely exchangeable between the end members enstatite, $MgSiO_3$, and ferrosilite, $Fe^{+2}SiO_3$. To represent this, the chemical formula of hypersthene is written $(Mg, Fe^{+2})SiO_3$. The replacement of one ion by another, maintaining charge balance in the structure, is facilitated by similarity in ionic radii. A second variety is called interstitial solid solution; this occurs where limited amounts of extraneous atoms can be accommodated in normally unoccupied structural sites. For instance, presence of minor amounts of the very small carbon atoms interstitial to the atoms of iron in iron metal give rise to the metallic phase known as steel (that is, carbon-bearing iron). The third variety of solid solution is termed omission solid solution. Basically, this type occurs where the presence of ions of variable valence or a change in the character of covalent bonding allows the periodic omission of atoms from structural sites. The iron sulfide pyrrhotite, FeS, exhibits such compositional variation because of systematic crystal defects, or omissions of iron atoms; accordingly, the formula for pyrrhotite may be written $Fe_{1-x}S$, where x has values between 0.0 and 0.2. Overall charge balance is maintained by the oxidation of Fe^{+2} and/or reduction of S^{-2}.

CRYSTALLOGRAPHY

Unit cell. No matter how complicated the geometrical relationship of the atoms, ions, or molecules in a mineral, its basic atomic structure exhibits a rigorous, continuous, three-dimensional periodicity. That is, if we were to start at any point in a structure and proceed along a straight line in any direction, within a few angstroms translation we would encounter an identical atomic arrangement. The distance connecting these two equivalent sites indicates the characteristic repeat periodicity in that direction. In general, repeats are different in different directions; however, a unique set of directions exists that forms a convenient triaxial reference system. These three principal directions—which, insofar as possible, coincide with symmetry elements of the structure (see definition, below)—are called the crystallographic axes. Unit repeats along these axes define a parallelopiped (a bricklike prism) that constitutes the basic building block of the structure—the unit cell.

Fourteen different parallelopipeds adequately describe the arrangements of unit repeats and crystallographic axes for all crystalline materials. A structure patterned after one of these parallelopipeds, or unit cells, may be considered as the fundamental block from which a specific crystalline material is constructed. In the architecture of any particular crystalline substance, only one type of building unit is used, and each is situated in an orientation and environment identical to those of its neighbors. As an example, construction of both a cube and an octahedron from cubic parallelopipeds is shown in figure 3.11. Of course, in a real crystal large enough to be seen by eye, billions of these unit cells are involved rather than the few hundred illustrated.

Symmetry. We have used the term symmetry before giving it a rigorous definition because the concept is intuitively obvious. The following definition embodies the major aspects of the concept. Symmetry is that property of an object whereby a specific operation (such as rotation) results in a position equivalent to the initial one for the object. Crystalline substances have planes and axes of symmetry, as well as other, more complicated types of symmetry; rotation about an axis or reflection through a mirror plane constitute some of the operations referred to above.

An axis of n-fold symmetry means that a rotation about the axis of 360 degrees divided by n results in equivalence. For example, a three-blade airplane propeller has threefold symmetry because a 120-degree rotation about the propeller shaft brings the blades into a position indistinguishable from—and identical too—the initial position. All objects, of course, have axes of onefold symmetry; near the other extreme, a right-circular cylinder, or cone, has an axis of infinite symmetry perpendicular to the base. Crystals, however, being constructed of rectilinear unit cells, display only one-, two-, three-, four-, and sixfold axes of symmetry—similar to the building blocks themselves. The geometric requirement is that the unit cells be stacked together in identical orientation leaving neither gaps nor overlaps. For a two-dimensional proof, you can try to arrange identical, regular polygons of either five-, seven-, or eightfold symmetry on a sheet of paper. You will soon discover what anyone who has ever installed floor tile knows; it is impossible to fill all the area with one of these shapes in identical orientations.

A plane of symmetry divides an object such as a crystal into two mirror images. Reflection rather than rotation is the symmetry operation here referred to. Most vehicles and animals, including humans (not flounders or oysters, however) display overall bilateral symmetry, and

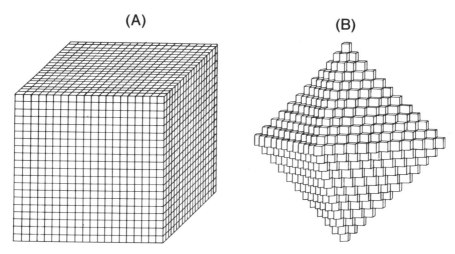

(A)　　　　　　　　　　**(B)**

FIGURE 3.11. Diagrammatic illustration of the architecture of crystalline materials from a particular type of unit cell (Ernst 1969). Shown are crystals whose external morphologies are (A) cube and (B) octahedron, both derived employing the same building block (cubic unit cell).

therefore a mirror plane, in their external morphology. Cars, trains, planes, and vertebrates lack internal bilateral symmetry due to functional constraints, a curious parallel between animate and inanimate objects.

The symmetry of crystals reflects the internal periodicity of all crystalline materials. The externally developed terminations that bound a perfectly developed, undistorted crystal may show more symmetry, but can exhibit no less symmetry, than the atomic structure constituting the building block from which the crystal is built.

PHYSICAL PROPERTIES OF MINERALS

Crystal Form. All crystal faces of the same type of shape and bearing a common spatial relationship to the crystallographic axes constitute one crystal form; they represent the collection of all symmetry-equivalent faces for a crystal. Where the crystals are undistorted, all such equivalent faces are identical in size and shape. Normally, however, due to differential growth of the faces, distorted crystals are produced. Even so, the angular relationships of the faces are preserved as a consequence of the symmetry of the unit cell building blocks. Figure 3.12 presents some common crystal forms: octahedron (eight faces); cube (six faces); dodecahedron (twelve faces); pyritohedron (twelve faces); rhombohedron (six faces); hexagonal prism (six faces); orthorhombric prisms (four

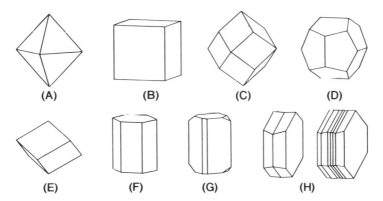

(A)   (B)   (C)   (D)

(E)   (F)   (G)   (H)

FIGURE 3.12. Examples of crystallographic forms: (A) octahedron; (B) cube; (C) dodecadron; (D) pyritohedron; (E) rhombohedron; (F) hexagonal prism (pinacoid); (G) combination of two orthohombic prisms (two pinacoids); (H) simply and multiply twinned low-symmetry crystals displaying four sets of pinacoids (Ernst 1969). All forms shown are perfect (undistorted).

faces each); pinacoids (two parallel faces); and singly and multiply twinned tablets (see below).

Twinning. Twinned crystals are composite grains that exhibit symmetrically related portions of contrasting structural orientations. Certain twins are united along a discrete reflection plane, or are related by an n-fold rotation. In neither case, however, does the twin mirror or twin axis coincide with a similar, intrinsic, pervasive symmetry element of the crystal structure. Twins may be simple (left-hand example of figure 3.12H) or multiple (right-hand example of figure 3.12H). Because of the re-entrant angles on some surfaces, multiply twinned crystals such as plagioclase may exhibit fine striations; the latter are helpful in identification of this mineral in hand specimen.

Cleavage. A regular (flat) fracture coincident in orientation with a possible crystal face is called a cleavage plane. Cleavages reflect planar weaknesses in a structure and usually are normal to directions of low bond densities and low bond strengths. Samples of a specific mineral species display the same cleavage because they all possess an identical internal arrangement of atoms, hence the same weakness directions. Cleavage planes should not be confused with crystal faces. Although it is quite true that both are structural planes, cleavage planes represent internal characteristics of a substance, and can be produced no matter how finely divided the sample becomes. On the other hand, crystal faces represent planar growth terminations, and once destroyed, they cannot be duplicated by finer division.

Fracture. Any surface of breakage that does not coincide with a possible crystallographic plane is termed a fracture. In a fracture, chemical bonds are broken in an irregular fashion unrelated to the symmetry of the internal structure. Quartz, SiO_2, for example, exhibits a curviplanar, or conchoidal, fracture, similar to that of broken glass. This type of breakage is distinguished readily from cleavage, because cleavage surfaces coincide with crystallographic planes, which are strictly planar, never curved or irregular.

Hardness. The resistance of a substance to abrasion is its hardness. Every mineral has a characteristic value that ultimately depends on bond strength. Different degrees of hardness may be determined by scratching one mineral by another. This operation actually breaks bonds and disrupts the atomic arrangement in a portion of the softer of the two minerals being tested. Table 3.2 presents an arbitrary but convenient numerical scale of relative hardness based on common minerals (Mohs'

TABLE 3.2. Mohs' Scale of Mineral Hardness

1	Talc	$Mg_3Si_4O_{10}(OH)_2$	SOFTEST
2	Gypsum	$CaSO_4 \cdot 2H_2O$	↑
3	Calcite	$CaCO_3$	
4	Fluorite	CaF_2	
5	Apatite	$Ca_5(PO_4)_3(OH,F,Cl)$	
6	Orthoclase	$KAlSi_3O_8$	
7	Quartz	SiO_2	
8	Topaz	$Al_2SiO_4(OH,F)_2$	
9	Corundum	Al_2O_3	↓
10	Diamond	C	HARDEST

scale, established by the Austrian mineralogist Frederick Mohs); the higher numbers here represent harder substances. Mineral hardness sets are not always available, so the resourceful or masochistic mineralogist may test an unknown material against a fingernail (hardness = 2–2½), a penny (hardness = 4), a tooth (hardness = 5), and a pocket knife or piece of glass (hardness = 5½–6).

Other Diagnostic Properties. Mineralogists employ other characteristics already discussed or intuitively familiar, such as specific gravity, color, streak (color of the powdered material), taste (solubility), and magnetic behavior, as an aid in identifying and understanding minerals and their occurrences.

MINERALOGY OF CARBON, SODIUM CHLORIDE, AND CALCIUM CARBONATE

Diamond and Graphite

Of all well-known polymorphs, those of carbon exhibit the most profound disparities in physical characteristics, reflecting major differences in atomic arrangements. The structures of diamond and graphite are compared in figure 3.13 as coordination-bonding models. The figure illustrates carbon coordination by minimizing the size of the atoms— represented by small black spheres—and by showing bond directions; neither graphite nor diamond structural model indicates the true packing of the carbon atoms. In both arrangements, carbon-carbon bonds are covalent; the electron clouds actually interpenetrate.

Carbon atoms are linked as perfectly planar hexagonal rings in the graphite structure. Each carbon is bonded to three other coplanar carbons. As shown in figure 3.13, these six-member rings are polymerized in two dimensions to yield an essentially infinite sheet, on an atomic scale. The distance between the centers of adjacent carbon atoms is 1.42 angstroms. The separation of successive sheets, however, is very large for minerals, 3.35 angstroms. Neighboring hexagonal sheets are offset in such a way that alternate carbons in one plane are bonded to noncorresponding atoms, three per six-member ring, in the next layer (see figure 3.13). This bond perpendicular to the sheets may be characterized as of the van der Waals type; a few resonating electrons flit episodically from layer to layer, providing a very weak attractive force binding the sheets together.

In the diamond structure, the carbon arrangement is somewhat similar to that of graphite in that it consists of "deformed," or puckered, hexagonal rings. Every other atom of a ring is displaced in one direction normal to the plane of the ring, whereas the other three are displaced in the opposite direction. An effect of this departure from strictly coplanar atoms is that in diamond, the carbon-carbon distances, 1.54 angstroms, slightly exceed those for carbon-carbon bonding in graphite within a strictly planar hexagonal ring. However, the displaced sets of atoms in diamond are juxtaposed with—and bonded to—carbon atoms of the adjacent puckered hexagonal rings, yielding an identical interatomic distance of 1.54 angstroms. Each carbon atom in diamond, then, is surrounded by four nearest neighbors, the centers of these surrounding atoms outlining a regular tetrahedron. Thus the polymerization of dia-

(A)   **(B)**

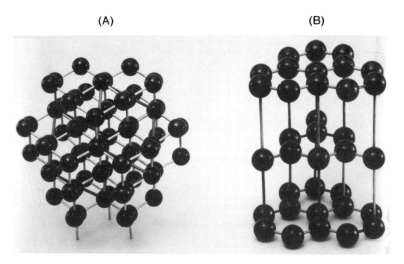

FIGURE 3.13. Coordination models of (A) diamond and (B) graphite (Ernst 1969). The weak van der Waals bonding at right angles to the hexagonal layering in the structure of graphite may be seen in (B).

TABLE 3.3. Physical Properties of Diamond and Graphite

Property	Diamond	Graphite
Unit parallelopiped	Cubic	Hexagonal
Common crystal form	Octahedron	Hexagonal platelets
Cleavage	Octahedral, perfect	Basal, perfect
Fracture	Conchoidal	None
Hardness	10	1–2
Color	Transparent colorless, colored, or opaque	Black
Streak	White	Black
Specific gravity	3.50	2.2
Special property	Formerly known as "a girl's best friend"	Soils fingers black

mond in three dimensions does not yield any structurally weak bond comparable to the interlayer, resonating-electron bond of graphite.

The contrasting physical properties of diamond and graphite are presented in table 3.3. The most common crystal form of diamond is the octahedron, whereas graphite occurs as hexagonal flakes and plates. Graphite has a perfect basal cleavage perpendicular to the van der Waals type bond and parallel to the sheets. Octahedral planes in diamond are coincident with the puckered hexagonal rings and have the highest concentrations of atoms; the number of covalent bonds normal to these planes is slightly less than that along other orientations, accounting for the octahedral cleavage. Diamond's great hardness is a consequence of the tightly knit polymerization of carbon and the exceptionally strong binding forces. In contrast, graphite has an extremely weak van der Waals–type bond normal to the hexagonal sheets; because this bond is readily broken, graphite is very soft. The covalent carbon-carbon bond distance within a graphite six-member ring is shorter than the corresponding internuclear separation in diamond, hence this within-ring attractive force is exceptionally strong; so when graphite is scratched, sheets are not disaggregated, they are merely translated laterally (i.e., shuffled about).

Graphite has a black color and streak, probably because weakly held electrons, resonating between successive sheets, are free to absorb light energy of any wavelength. In diamond, on the other hand, electrons are firmly restricted to specific orbital configurations and energy levels, so electromagnetic radiation suffers little absorption and passes through the crystal with almost no attenuation by absorption. However, the presence of trace amounts of impurities in some diamonds results in color tints or even opacity. The specific gravities of these two polymorphs reflect their very different atomic arrangements, quite compact in diamond and much more open in graphite (table 3.3).

Although diamond is a very rare mineral, it has been found in stream gravels, especially in India, Brazil, the Union of South Africa, and Zaire. In fact, most diamonds used industrially are mined from such placer deposits. The primary occurrence of diamonds in eastern Siberia, south Africa, and western Australia is in the so-called kimberlite pipes, which are funnel-, pipe-, or carrot-shaped bodies of a type of mantle-derived igneous rock known as peridotite (see chapter 4). Kimberlites are rich in MgO, FeO, and CaO, uncommonly low in SiO_2 relative to other igneous rocks, and partly altered to hydrous, calcareous mineral assemblages. Having originated at considerable depths, they apparently have been intruded into upper portions of the crust as a moderately hot, fluidized mush of solids, propelled by an expanding CO_2-rich aqueous volatile phase.

The temperature-pressure diagram for the stable polymorphs of carbon is shown in figure 3.14. Because diamond has a smaller volume than graphite, its formation is favored by high pressures. As is evident from this diagram, diamond-bearing peridotites must have originated at depths exceeding about 125 kilometers, where the weight of the overlying column of rocks or lithostatic pressure is approximately forty kilobars or more, provided the geothermal gradient shown is correct. The geothermal gradient is a P-T trajectory—that is, a line on a pressure-tempera-

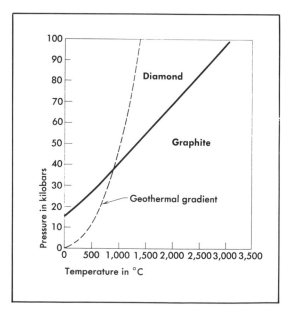

FIGURE 3.14. Phase relations for the polymorphs of carbon (Ernst 1969).

ture diagram—indicating the temperature within the Earth as a function of depth. Evidently, diamond pipes contain material that is derived from well within the mantle, near or at the base of relatively thick lithosphere.

Within the continental crust, where confining pressure approaches a maximum of about ten kilobars, graphite is the stable polymorph of carbon under all P-T conditions, as clear from examination of figure 3.14. Diamond may persist for great lengths of time in crustal regions, and in the uppermost mantle as well, because the transformation to graphite proceeds very slowly below 1000 degrees Celsius. Some form of carbon, derived from decomposed organic material, is a minor but virtually ubiquitous constituent of sedimentary rocks and of recrystallized (metamorphic) rocks derived from sediments. The biogenic carbon is amorphous in most sediments but recrystallizes by degrees to graphite during metamorphism. Under certain conditions of high-temperature metamorphism, the graphite is progressively oxidized to carbon dioxide, which in turn is driven off, leaving behind a more reduced assemblage of recrystallized solids.

Halite

Unlike diamond and graphite, halite, NaCl, is a compound; it is a rather simple example of the type of structures displayed by the rock-forming minerals to be discussed later on. Earlier, it was pointed out that, in halite, each sodium ion is surrounded by six chlorine ions and conversely each chlorine ion by six sodium ions. The ionic radii of Na^+ and Cl^- are approximately 1.02 angstroms and 1.81 angstroms, respectively; the radius ratio is therefore 0.56, which accounts for the octahedral coordination (see figure 3.9). The crystal structure, shown in figure 3.15, is based on a cubic unit cell. Local electrostatic stability is maintained by the nondirectional distribution of attractive forces. Molecules as discrete entities do not exist in this mineral structure, nor in most others. Bonding has a markedly ionic character, as is evident from the electronegativity differences illustrated previously in figure 3.8.

The physical properties of halite are listed in table 3.4. Cubes represent the most common crystal form. The low hardness of NaCl reflects the relatively weak attractive forces between large, monovalent ions. Three mutually perpendicular cleavage planes are a consequence of the high atomic populations in planes parallel to the faces of the unit cell. The number of bonds normal to these layers is slightly less than the number of bonds perpendicular to other structural planes. The lack of free electrons accounts for halite's transparency, because the various wavelengths of incoming light are not differentially absorbed. Tinted varieties of NaCl result from the presence of light-absorbing impurities or are due to ionic omissions (defects) that produce unsatisfied electron orbitals. Strong polarization, or distortion of the electron clouds surrounding the ions, would allow tight packing, but this does not occur in halite due to the large sizes and small ionic charges. Packing is essentially that of incompressible spheres and, inasmuch as the atomic species have fairly small masses, specific gravity is rather low.

In nature, minor amounts of NaCl are precipitated from cooling hydrothermal solutions. However, halite is most abundant in sedimentary rocks, where it occurs as a chemical precipatate from evaporating seawater. Because NaCl is highly soluble in H_2O, large amounts of solution must be dehydrated before the brines become concentrated enough to precipitate halite. Such situations arise in continental lakes and constricted marine embayments where evaporation matches the inflow of fresh water (containing minor dissolved salts) or exceeds the rate of homogenization with normal seawater. Current examples of continental evaporite deposits include Great Salt Lake, Utah, the Salton Sea, California, and the Dead Sea between Israel and Jordan. In marine environments, salt deposits are accumulating around the southeast edge of the Mediterranean, and along margins of the Persian Gulf and the Red Sea. The presence of ancient evaporite deposits in the rock record demonstrates that similar conditions were perhaps equally prevalent in the past. Although the mean sodium chloride content of seawater probably

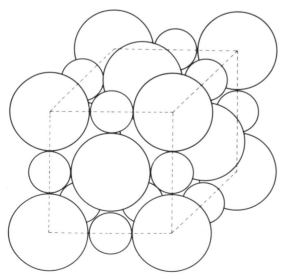

FIGURE 3.15. Packing model of the halite structure (Ernst 1969). The large spheres represent Cl^-, the small spheres Na^+. The unit cell is outlined by dashed lines.

TABLE 3.4. Physical Properties of Halite

Unit parallelopiped	Cubic
Common crystal form	Cube
Cleavage	Cubic, perfect
Fracture	None
Hardness	2.5
Color	Transparent colorless, pale-tinted
Streak	White
Specific gravity	2.16
Special property	Salty taste

has fluctuated over geologic time, there seems to be no compelling reason to believe that the oceans are becoming progressively saltier with the passage of time.

Calcite and Aragonite

The compound calcium carbonate, $CaCO_3$, crystallizes in nature as two different polymorphs, calcite and aragonite. Schematic coordination models of the structures, and the packing configuration of $(CO_3)^{-2}$, are illustrated in figure 3.16. Both crystal structures contain negatively charged planar $(CO_3)^{-2}$ groups, as do all carbonates; the oxygens of these carbonate radicals are cross-linked to calcium ions. The central carbon atom is coordinated to three nearest-neighbor oxygens by strong, partially covalent bonds, and the over-all charge on the triangular anion complex is -2. In both calcite and aragonite, corresponding $(CO_3)^{-2}$ groups of adjacent structural planes are opposed 180 degrees in orientation. Although these triangular groups exhibit marked electron-sharing, they behave as anion complexes; each of the constituent oxygens is bonded ionically to two CA^{+2} cations in calcite, and three Ca^{+2} cations in aragonite. Each $(CO_3)^{-2}$ group is surrounded by six cations, and vice versa. Calcite has a relatively open structure compared with the denser atomic packing of aragonite.

The physical properties of calcite and aragonite are compared in table 3.5. Calcite crystallizes with many different morphologies, but rhombohedra (figure 3.12E) are common. The most typical aragonite crystals are prisms. The rhombohedral and prismatic cleavages of calcite and aragonite, respectively, are parallel to structural planes of maximum atomic density; hence the number of bonds normal to these planes is moderately low. The bonds broken are those between Ca^{+2} and $(CO_3)^{-2}$ groups rather than the stronger bonds that link carbon and oxygen in the planar anion complexes. Aragonite is slightly harder than calcite because the larger number of calcium-oxygen bonds and more compact arrangement lead to slightly greater total attractive forces between cations and anion groups. This compact arrangement also explains the higher specific gravity of aragonite compared with calcite. The marked electron-sharing and electron-donating characteristics of the attractive forces mean that no weakly bound electrons are available to absorb electromagnetic radiation; in the absence of impurities, $CaCO_3$ polymorphs are therefore transparent. The application of cold, dilute hydrochloric acid to calcite or aragonite causes "fizzing"; that is, CO_2 gas is liberated—a diagnostic test for carbonate.

The phase diagram for $CaCO_3$ is presented in figure 3.17. Aragonite, which has a more compact (smaller volume) atomic arrangement than calcite, is stable on the high-pressure, low-temperature side of the P-T curve relating the two polymorphs. With an independent means to determine temperature, this curve allows the estimation of the minimum or maximum pressure that attended equilibrium crystallization of aragonite- or calcite-bearing rocks, respectively.

Because many divalent cations have ionic radii approaching that of Ca^{+2}, (see table 3.1), the corresponding carbonates of these cations crystallize with calcite- and aragonite-type structures. Curiously, although

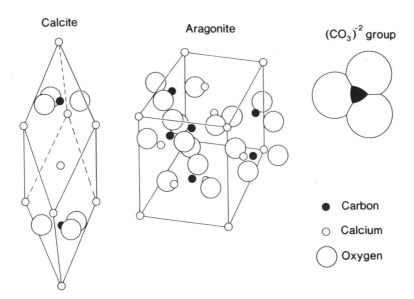

FIGURE 3.16. Schematic arrangements of Ca^{+2} and $(CO_3)^{-2}$ groups in calcite and aragonite (Ernst 1969). The unit cell is outlined by solid lines.

TABLE 3.5. Physical Properties of Calcite and Aragonite

Property	Calcite	Aragonite
Unit parallelopiped	Trigonal (see figure 3.16)	Orthorhombic (see figure 3.16)
Common crystal form	Rhombohedron, many others	Prism
Cleavage	Rhombohedral, perfect	Prismatic, distinct
Fracture	None	Subconchoidal
Hardness	3	3.5–4
Color	Transparent colorless, colored	Transparent colorless, colored
Streak	White	White
Specific gravity	2.71	2.94
Special property	Dilute HCl causes effervescence	Dilute HCl causes effervescence

calcite displays a more open configuration of atoms than aragonite, the mean size of the cation position is smaller. For this reason sixfold coordination cations of ionic radii less than that of Ca^{+2}, such as Mg^{+2}, Fe^{+2}, Zn^{+2}, and Mn^{+2}, crystallize with the calcite-type crystal structure, whereas those with ionic radii exceeding calcium, for example, Sr^{+2} and Ba^{+2}, possess the atomic configuration of aragonite.

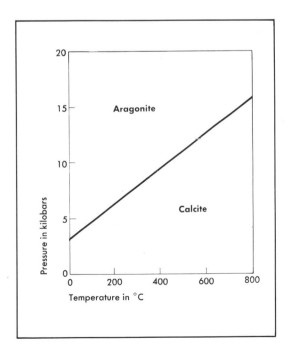

FIGURE 3.17. Phase relations for the polymorphs of calcium carbonate (Ernst 1969).

Carbonates are important rock-forming minerals, and their occurrences are diverse. In metamorphic rocks, calcite-type carbonates occur as essential or accessory minerals and as vein and fissure fillings; rarely, aragonite occurs in high-pressure, low-temperature metamorphic rocks. Certain peculiar igneous rocks, known as carbonatites, have been described, but carbonate minerals are not important in most igneous occurrence. Carbonates are most abundant, however, in sedimentary rocks where they are present as mechanically transported and deposited grains, as a cementing agent in sandstones and mudstones, and as the principal mineral in chemically precipitated carbonate rocks, especially limestone and dolomite. Shell material consists mainly of $CaCO_3$, so biogenic carbonate is present in many sediments.

As illustrated in figure 3.17, aragonite is unstable near the Earth's surface. It is being deposited, however, on the Bahama Banks and in caves, as well as being secreted by certain invertebrates. Such occurrences can be accounted for by local supersaturation of carbonated solutions. Eventually, the aragonite so produced spontaneously transforms to the stable low-pressure equivalent, calcite.

___ A BRIEF INTRODUCTION TO THE MINERALOGY OF THE SILICATES

As was pointed out in chapter 1, the Earth's crust and upper mantle consist chiefly of silicates. Rather than describing these extremely important minerals in detail—their structures and chemical variabilities are quite complex—we will classify and discuss the general characteristics of this structural diversity, then briefly mention some of the physical properties and chemistries of the contrasting mineral groups. The silicates all possess rather intricate structures, so only simplified descriptions will be presented.

One of the important cations in silicates obviously must be silicon. This small, highly charged positive ion, which is quadrivalent and has an ionic radius of 0.26 angstroms or slightly more, is surrounded by four much larger nearest-neighbor oxygens. Each oxygen carries a divalent negative charge and has an ionic radius of 1.35 angstroms. The radius ratio is 0.20, which more or less explains the tetrahedral coordination, as shown in figure 3.9. A net negative charge on the resulting complex, $(SiO_4)^{-4}$ necessitates bonding to additional silicons and/or other cations. The degree of tetrahedral polymerization (or linkage) provides the basis for a structural classification of the silicates (see figures 3.18 through 3.21). In the following discussion, for simplicity, we will consider bonding to be ionic, but as is plain from figure 3.8, this assumption is only approximately correct.

In some minerals, such as olivines and garnets, individual silicon tetrahedra share oxygens only with cations other than silicon. Their chemical formulas therefore include $(SiO_4)^{-4}$ anion complexes. A typical olivine, forsterite, is written Mg_2SiO_4, and almandine garnet is $Fe_3^{+2}Al_2(SiO_4)_3$. Each of these mineral groups shows extensive solid solution. Common olivines [$(Mg, Fe^{+2})_2 SiO_4$], for instance, range in composition between forsterite and the ferrous iron end member fayalite, $Fe_2^{2+}SiO_4$. The silicon-oxygen arrangement in an independent tetrahedron is illustrated in figure 3.18A.

In epidote, two silicon tetrahedra share a common oxygen; for such paired tetrahedra, the silicon-oxygen ratio is not 2:8, but 2:7. The shared anion is known as a bridging oxygen because it cross-links two tetrahe-

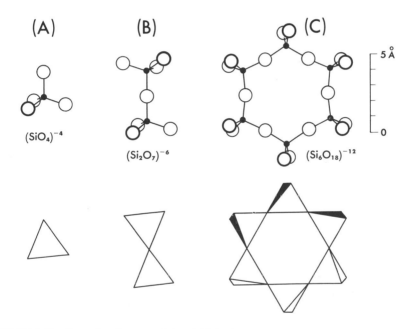

FIGURE 3.18. Coordination models of (A) independent silicon tetrahedra such as occur in the olivines and garnets; (B) paired tetrahedra as found in the epidote minerals; and (C) six-member rings characteristic of the mineral tourmaline and beryl (Ernst 1969). In this and the subsequent diagrams of figure 3.19, 3.20, and 3.21, small spheres represent silicon (and minor aluminum in some cases); large spheres represent oxygen. In addition to the coordination models, arrangements of silicons and oxygens are shown as individual tetrahedra. Because of strong bonds in all directions, minerals characterized by isolated, paired, and ringed silicon-oxygen tetrahedra are hard, refractory and do not in general possess good cleavages.

drally coordinated silicon atoms. The valence orbital of the bridging oxygen is completed by the statistical donation of an electron from each of the two nearest-neighbor silicons. The other tetrahedrally disposed oxygens each receive one electron from their central silicons, but must obtain the second electron from peripheral, coordinating cations. Because there are six such nonbridging oxygens, the anion complex has a net negative charge of six, $(Si_2O_7)^{-6}$. Paired silicon tetrahedra are illustrated in figure 3.18B. The structural formula of epidote, $Ca_2Fe^{+3}Al_2O$ $(OH)(Si_2O_7)(SiO_4)$, indicates that this complex hydrous silicate contains paired as well as isolated silicon-oxygen tetrahedra.

Increased polymerization results in three-, four-, or six-member rings of tetrahedra. Here, two oxygens in each tetrahedron are bridging oxygens, and the remaining two must each statistically obtain an electron from other cations. Thus, the net negative charge on three-, four-, and six-member rings is six, eight, and twelve respectively. The arrangement in a six-member ring is shown in figure 3.18C. The $(Si_6O_{18})^{-12}$ group indicating six silicon tetrahedra per ring, characterizes the mineral beryl, $Be_3Al_2Si_6O_{18}$.

More extensive polymerization of silicon tetrahedra gives rise to chains, sheets, and frameworks. The two fundamental types of silicon-oxygen chains, single and double, are illustrated in figure 3.19. While triple chains are possible (and do occur rarely), sheet silicates of the same composition appear to be more stable. In single-chains silicates, as figure 3.19A shows, two of the oxygens in each tetrahedron are bridging oxygens. Only half of each bridging oxygen may be assigned to a specific tetrahedron, and the silicon-oxygen ratio reduces from 1:4 in the case of isolated tetrahedra to 1:3 in single chains. Every tetrahedron contains two nonbridging oxygens, each of which receives an electron from a peripheral cation. As in ring silicates, the basic single-chain unit, then, is $(SiO_3)^{-2}$, which is typical of the pyroxene mineral group. Examples include enstatite, $MgSiO_3$, and diopside, $CaMg(SiO_3)_2$. The former is an example of an orthopyroxene (three mutually perpendicular crystallographic axes), the latter is a clinopyroxene (one crystallographic axis inclined to the plane defined by the orthogonal intersection of the other two).

The most important double-chain silicates belong to the amphibole group. The structure consists of two single chains of silicon-oxygen tetrahedra with alternate tetrahedra cross-linked to the adjacent chain, as shown in figure 3.19B. Within the chain repeat of four silicons, five bridging oxygens are shared between two tetrahedra; hence there are only eleven instead of sixteen oxygens in the unit repeat, which is written $(Si_4O_{11})^{-6}$. The net negative charge is six, because there are six nonbridging oxygens per four silicons. Typical double-chain silicates include the amphiboles anthophyllite, $Mg_7(Si_4O_{11})_2(OH)_2$, and tremolite, $Ca_2Mg_5(Si_4O_{11})_2(OH)_2$. Much more complicated solid solutions include the common amphibole, hornblende, $(K,Na,Ca)_{2-3}(Mg,Fe^{+2},Fe^{+3},Al)_5$ $(Si,Al)_8O_{22}(OH)_2$.

Amphiboles and pyroxenes display essentially infinite silicon tetrahedral polymerization in the direction parallel to the chain length. In the sheet silicates, cross-linkage is carried further, and the tetrahedral units are extended in two dimensions to produce layers, as illustrated in figure 3.20. Just as double chains may be described as six-member tetrahedral rings polymerized in one direction, sheet structures may be described as six-member tetrahedral rings cross-linked in a plane (com-

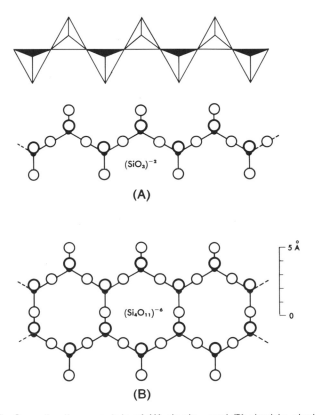

FIGURE 3.19. Coordination models of (A) single- and (B) double-chain polymerized silicons and oxygens; linked tetrahedra are also illustrated in (A) (Ernst 1969). These linear structural units occur in (A) pyroxenes and (B) amphiboles, respectively. Two prismatic cleavages that avoid breaking the bridging oxygen bonds characterize chain silicates. Because of high bond strengths, these minerals are hard and tend to have high melting points.

pare figures 3.19B and 3.20). In sheet silicates, three of the four oxygens surrounding a central silicon are bridging oxygens. Therefore, per silicon, there is one nonbridging oxygen, and the equivalent of one and a half oxygens whose charge requirements are completely satisfied by the tetrahedrally coordinated cations. The basic repeat unit includes four silicons and ten oxygens, so the formula of the tetrahedral unit is $(Si_4O_{10})^{-4}$.

Talc and muscovite mica are sheet silicates, as indicated by structural formulas $Mg_3(Si_4O_{10})(OH)_2$ and $KAl_2(Si_3AlO_{10})(OH)_2$, respectively. They are colorless, whereas the presence of ferrous iron accounts for the appearance of the black mica, biotite, $K(Mg,Fe^{+2})_3(Si_3AlO_{10})(OH)_2$. Clay minerals and chlorite, which are important in sedimentary and metamorphic rocks, respectively, also constitute major groups of sheet silicates. Although we have implicitly assumed that all tetrahedrally coordinated cations are silicon, this is not strictly true for many rock-forming silicates, as illustrated by the muscovite and biotite formulas just pre-

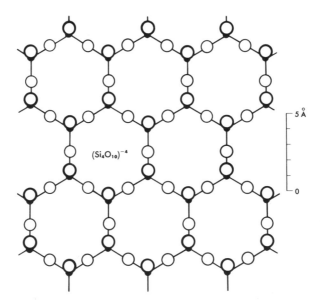

FIGURE 3.20. Coordination model of a tetrahedrally polymerized sheet (Ernst 1969). Perfect cleavage parallel to the silicon-oxygen tetrahedral layers totally avoids the bridging oxygens. These planar structural units characterize the micas, chlorites, and clay minerals, known collectively as layer silicates. Weak bonding across the sheets causes many sheet silicates to decompose readily at elevated temperatures.

sented. In micas, one fourth of the tetrahedral sites are occupied by aluminum and three fourths contain silicon. Thus, the basic structural unit becomes $(Si_3AlO_{10})^{-5}$, rather than $(Si_4O_{10})^{-4}$ as in talc. Certain amphiboles such as hornblende, and some pyroxenes, too, contain aluminum in fourfold coordination. As we will now see, tetrahedrally coordinated aluminum is absolutely essential in the feldspars.

Feldspars and quartz are the most abundant minerals in the Earth's continental crust. In these mineral groups, the tetrahedral units are polymerized in three dimensions, so each and every oxygen is shared between two silicons (or one silicon and one aluminum). This three-dimensional linkage results in a tetrahedral framework analogous to the one shown in figure 3.21, so minerals possessing this type of configuration are called framework silicates. For quartz and other polymorphs of SiO_2, silicon is the only tetrahedrally coordinated cation. It is sur-

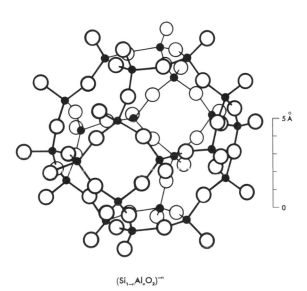

$(Si_{1-n}Al_nO_2)^{-n}$

FIGURE 3.21. Coordination model of a three-dimensional framework of polymerized tetrahedra (Ernst 1969). Such minerals, known collectively as framework silicates, in general have imperfect or no cleavage because of strong bonds in all directions. They are relatively hard and refractory, except for species that contain hydroxyl, $(OH)^-$, and/or water of hydration. Their open atomic packing imparts low specific gravities to most minerals of this structural type. The example illustrated is a complex hydrous framework silicate of the zeolite group, but other common phases such as quartz and feldspars have somewhat similar, although generally more compact structures.

rounded by four bridging oxygens (each counted as belonging one half to each of two tetrahedra), so the structural formula reduces to SiO_2. In feldspar structures, however, between one fourth and one half of the tetrahedral sites contain aluminum. Accordingly, the framework is represented by portions of mineral formulas such as $(AlSi_3O_8)^{-1}$ and $(Al_2Si_2O_8)^{-2}$. The substitution of Al^{+3} for Si^{+4} generates a charge imbalance that must be compensated for by an equivalent number of monovalent cations or half as many divalent cations.

Formulas of typical framework silicates, the plagioclase feldspars, are albite (abbreviated Ab), $NaAlSi_3O_8$; and anorthite (abbreviated An), $CaAl_2Si_2O_8$. They form a complete solid-solution series. For our purposes, we will distinguish three compositional ranges: sodic (less than 10 percent anorthite); intermediate (10–50 percent anorthite); and calcic (greater than 50 percent anorthite). Yet a third type of feldspar needs to be noted. In the K-feldspar (abbreviated Or), $KAlSi_3O_8$, orthoclase, potassium resides in the cation site occupied by sodium in albite. The alkali feldspars represent solid solution between $NaAlSi_3O_8$ and $KAlSi_3O_8$. The so-called ternary feldspars, therefore, consist of solid solutions among three end members, anorthite, albite, and orthoclase; only two essentially binary solid solutions are important, anorthite-albite (plagioclase series), and albite-orthoclase (alkali feldspar series).

With the exception of the silica polymorphs of nearly invariant SiO_2 composition, extensive substitutions in the different cation sites are responsible for the broad range of compositions found among most rock-forming silicates. In spite of these substitutions, the crystal structures of a particular silicate mineral group are rather uniform in their atomic arrangements because of high silicon-oxygen (and aluminum-oxygen) bond strengths. Therefore, the type of tetrahedral polymerization controls the physical properties of the individual silicate groups to a very large extent, regardless of chemical variations. For instance micas, no matter what their chemistry, all exhibit a perfect basal cleavage parallel to the silicon-oxygen sheets; amphiboles and pyroxenes, in contrast, possess two excellent prismatic cleavages, the intersections of which define the direction of tetrahedral polymerization or chain extension.

The occurrences of the silicates are so diverse as to defy simple treatment. Quartz and the sodic plus potassic feldspars are the most important minerals of the continental crust and are major phases in nearly all principal rock types; sheet silicates are widespread but slightly less abundant. Chain silicates and minerals characterized by structures containing isolated tetrahedra are generally confined to the more refractory (stable to high temperatures) igneous and metamorphic rocks of the continental crust. The oceanic crust consists chiefly of calcic plagioclase, pyroxenes, and olivine, in that order of abundance. The uppermost mantle is rich in magnesian olivine, and contains lesser amount of magnesium ortho- and clinopyroxenes; at greater depths, simple oxide structures apparently prevail in the lower mantle. Occurrences of silicate minerals will be noted more systematically in the mineralogic description of crustal rocks, to be presented in chapter 4.

$\equiv 4$
$\equiv$ ROCKS: MATERIALS OF THE
$\equiv$ LITHOSPHERE

As previously mentioned, the earth materials that constitute both continental and oceanic crusts are of three principal types: (1) igneous— rocks that have solidified from a melt (magma); (2) sedimentary—rocks produced by accumulation of solids that have settled out of a transporting medium such as wind, ice, or, most commonly, water; and (3) metamorphic—rocks that have been formed by chiefly solid-state changes due to recrystallization and/or deformation of pre-existing rocks.

By far the dominant process responsible for generation of the crust has involved partial fusion (i.e., incomplete melting) of the more deeply buried, hence hotter, mantle, separation of the molten material from the refractory (high thermal stability) crystalline residuum, and subsequent bouyant rise of the magmas so produced toward the surface. Because this is the main process through which the crust has formed, such igneous products are referred to as primary rocks. The mechanism of partial melting, collection, and bouyant ascent of the silicate liquid, and solidification in the crust, is a long-continued one that apparently is also accompanied by devolatilization (removal of gases—outgassing) of the mantle. Due to the Earth's substantial internal heat content and consequent high temperatures, the crust, hydrosphere, and atmosphere have accumulated and evolved over the course of geologic time. A scenario that describes the chemical differentiation of our planet will be presented in chapter 7.

Surficial processes involving the erosional agents wind, water, and ice are attended by chemical interactions, dissolution, and mechanical abrasion of the crust; these result in the disaggregation/decomposition and transportation of pre-existing rocky materials, and deposition as sedimentary layers (strata). Later burial and folding of these and other precursor units give rise to the deformation and mineralogic reconstitution characteristic of metamorphic rocks. Thus, both sedimentary and metamorphic lithologies are secondary in nature, because they require the prior existence of an earlier rock unit; accordingly, only igneous rocks can be considered as the primary materials of the Earth's outer rind—i.e., derived directly from the mantle. Of course, some magmatic rocks are secondary too, such as those that represent the partial fusion of deeply buried, pre-existing crustal rocks (igneous, metamorphic, and/ or sedimentary). It is clear, therefore, that the crust must bear witness somehow to complicated and redundant chemical and physical processes taking place throughout the lithosphere and in the deeper upper mantle in response to the Earth's present and past thermal regime. This subject will be returned to at the end of this chapter where the rock cycle is discussed.

IGNEOUS ROCKS

Occurrence of Igneous Rocks

These lithologies constitute approximately 75 percent of the continental crust, and well over 90 percent of the oceanic crust. Some have been derived directly from the mantle, whereas others represent partly remelted oceanic or continental crustal materials. The high temperatures required to melt igneous rocks in the laboratory and which have been measured for erupting lavas (commonly in excess of 800–1100 degrees Celsius, depending on the composition of the material and the attendant pressure), demonstrate that magmas must originate at considerable depths in the Earth, where such temperatures are characteristic.

Igneous rocks exhibit rather limited compositional variations. Their principle constituent oxide is silica, SiO_2, which ranges from about 45 percent to 75 percent by weight in the common types. Where the silica content is low, magnesium and iron oxides usually are major components. This is reflected by the abundance of dark magnesium- and iron-bearing minerals such as olivine, the pyroxenes, the amphiboles, and biotite. Silica-poor igneous rocks are called subsilicic; ferromagnesian-rich rocks are termed mafic. Generally speaking, most subsilicic rocks are mafic and vice versa. The yet more magnesium- and iron-rich, silicon-, sodium-, and potassium-poor rocks characteristic of the upper mantle are known as ultramafic rocks. In igneous rocks in which the silica contents exceed about 60 percent to 65 percent by weight, the mineral quartz occurs and is associated with potassium- and sodium-rich feldspars with or without muscovite, and only minor amounts of dark, ferromagnesian minerals. Such rocks are light-colored and are termed silicic and/or felsic (i.e., feldspar-rich). Most felsic igneous rocks are silicic, and contrariwise. The color index of a rock is defined as the volume percentage of dark, or ferromagnesian, minerals: the lower the color index, the more felsic and silicic the rock.

Two principal types of igneous rocks occur, extrusive and intrusive. To the first group belong those igneous rocks that have reached the surface of the solid Earth in a largely molten condition. Lava flows, for instance, are streams of magma poured out on the surface, and volcanic ash is magma that has been blown apart during extrusion by the explosive expansion of dissolved gases as external pressure (due to the overlying magma column) is reduced. Volcanic activity is among the most spectacular and catastrophic of geologic processes and rivals earthquakes in terms of devastation and loss of life. A vivid example, illustrated in figure 4.1, is the 1963 birth and growth of Surtsey, a volcanic island just south of Iceland. It is a portion of the Mid-Atlantic Ridge that rose above sea level due to repeated eruptions. Intrusive igneous rocks are those that crystallized from magmas that did not reach the surface. Such rocks cool more slowly than extrusives, and they retain their dissolved volatile constituents more completely. Hence, intrusive rocks contain coarser mineral grains and a higher proportion of hydrous mineral phases than do the finer-grained extrusive rocks.

Two broadly contrasting modes of volcanic activity have been recognized: (1) the quiescent, or fissure, type; and (2), the more violent central-eruptive type. The great volumes of mafic lava that have poured out on the Earth's surface episodically through geologic time have issued principally from fissures, along which magma from considerable depth has gained access to the surface. Lavas are dominantly dark, subsilicic basalts, and they are characteristically quite fluid. On the continents they accumulate to form broad plateaus, approaching a kilometer or more in thickness and tens of thousands of square kilometers in area. Major occurrences include the Deccan traps of the Indian Peninsula; the Columbia and Snake River plateaus of Washington, Oregon, and Idaho; the Tertiary volcanics of Britain, Iceland, and Greenland; and the Karroo basalts of South Africa. Even more extensive are the basalts (and gabbros, their intrusive equivalents) that originate at spreading ridges and floor the world's ocean basins. These accumulations average approximately five kilometers in thickness, as determined by seismic measurements (see chapter 1).

More familiar but volumetrically less important are extrusive rocks of the central-eruptive type, which build distinctive volcanic cones. Two major varieties can be distinguished: shield volcanoes and stratovolca-

FIGURE 4.1. Growth of the basaltic island Surtsey along the Mid-Atlantic Ridge just south of Iceland. (A) After earlier submarine construction of the volcanic edifice, subaerial eruption commenced on November 14, 1963. (B) Appearance of the island on April 20, 1964; lava flows are advancing toward the foreground and have reached the sea on the left. (From Thorarinsson 1964, plates 1 and 27.)

(A)

B)

noes. To the former group belong the massive, broad volcanoes with a shield-like outline and gently sloping flanks, such as Mauna Loa, Mauna Kea, and Kilauea in Hawaii, as presented in figure 4.2. Such structures are produced by the episodic extrusion of highly fluid, subsilicic lava from a central conduit; the low viscosity promotes the formation of a broad, gently sloped, but in some cases, enormous central cone. In contrast, steep-sided stratovolcanoes are built up through the eruption of more viscous lavas having compositions intermediate between mafic and felsic melts; explosive outbursts of ash are also common to this type of volcanism. Famously photogenic cones such as Mounts Fuji, shown in figure 4.3, Mayon, Vesuvius and Rainier (of Japan, the Philippines, Italy, and Washington State, respectively) are of this type. Both shield and stratovolcanoes commonly exhibit flank eruptions and build satellite cones because of through-going faults and fractures. In many cases, in fact, active central-eruptive type volcanoes themselves are disposed at regular intervals along linear trends—evidently marking deep regional fissure systems along which magma has risen towards the surface from great depth.

The mafic, subsilicic lavas tend to be relatively thin, widespread flows, whereas the more silicic, felsic flows are rather thick and topographically prominent, reflecting the greater viscosity of the latter type. Fluidity depends on both the temperature of eruption and the composition. Subsilicic lavas are extruded at temperatures exceeding 1100 de-

FIGURE 4.2. The immense Hawaiian shield volcano Mauna Loa, with a summit elevation 4,170 meters (13,680 feet) above sea level, though topographically indistinct, dominates the skyline. In the foreground, note that a lava lake occupies a breached crater pit, Kilauea Iki, a satellite of Kilauea volcano. (Photo by author, 1965.) This igneous complex, located in a midplate regime, is a reflection of the passage of Pacific lithosphere over a fixed hot spot in the deep upper mantle (see chapter 2, and figure 5.4).

grees Celsius, and so are less viscous than the lower-temperature silica-rich lavas. Moreover, the greater the amount of SiO_2 per unit mass, the more extensive the atomic-scale polymerization, or linkage of silicon-oxygen tetrahedral groups in the melt, resulting in greater stiffness of the more siliceous liquids.

Depending on their viscosity, lavas are ropy or blocky, or they may be massive. Individual flows range from less than a meter to several hundred meters in thickness. Finer-grained margins of lava flows are developed by rapid cooling at contacts with older rocks, with the atmosphere, or with bodies of water. Polygonal sets of fractures, known as columnar joints, develop at right angles to the contacts or cooling surfaces, especially in the finer-grained, chilled portions of flows. Commonly, the tops of flows are to some extent oxidized through reaction with air while the lavas are still hot. In some cases, subaqueous extrusion results in the coagulation of ellipsoidal, meter-size blobs of lava that resemble pillows—hence the term pillow lavas. These features are best developed in oceanic basalts.

Dissolved volatiles (mostly H_2O, to a lesser extent CO_2) can escape

FIGURE 4.3. The nearly symmetrical central cone of Mount Fuji, seen from the southwest; elevation of the summit is 3,778 meters (12,395 feet) above sea level. (Photo by author, 1963.) Similar to other northeastern Japanese stratovolcanoes, Fuji is sited over the northwestward-subducting Pacific lithospheric plate.

easily into the atmosphere from highly fluid, subsilicic lavas upon extrusion, but the expansion of volatile constituents generally causes relatively viscous, more silicic melt to be transformed into a light, frothy aggregate termed pumice, or in some instances to shatter explosively. In the latter case, the partly molten particles are blown out of the volcanic throat and are rapidly quenched on contact with the atmosphere. Eruptions of this sort result in the accumulation of volcanic debris and ash, principally downwind from the source. Such deposits are called pyroclastics (that is, fiery fragments), and were characteristic of the Mount St. Helens eruption in 1980 (see chapter 5 for a description of this event).

Three kinds of volcanic ejecta are distinguished on the basis of grain size. The finest debris, less than four millimeters in particle size, is termed ash. Pea- to walnut-sized cinders ranging from four to thirty-two millimeters in diameter are called lapilli. The fragments that exceed thirty-two millimeters in diameter are termed blocks if angular, or bombs if partially rounded; most of the former may represent ripped-off fragments of the volcanic conduit, whereas the latter probably reflect the ejection and flight of partly congealed lava. Once they are consolidated, pyroclastic deposits are termed tuff, lapilli tuff, tuff breccia (containing blocks) and agglomerate (containing bombs).

Pompeii, Italy, was buried by ash in A.D. 79 during a pyroclastic eruption of Vesuvius. Nearby Herculaneum was wiped out by mudflows consisting of rainsoaked ash newly deposited on the unstable, oversteepened flanks of the volcano. In another kind of eruption, hot, incandescent ash issues from the volcanic throat and accumulates on the cone flanks while still emitting gassy constituents. This material, being well lubricated by the gas, can move downslope at very high velocity as a nuée ardente (glowing cloud). Such an eruption of Mount Pelée obliterated St. Pierre, Martinique, in 1902. These are but two of numerous pyroclastic eruptions that have caused catastrophic loss of life. The Mount Pelée disaster alone killed between 25,000 and 40,000 people!

Being hidden from view, deep-seated igneous processes are much less obvious than the various kinds of volcanic activity. Based on their geometries, three principal types of intrusive igneous bodies can be distinguished: subjacent masses (downward expanding); tabular bodies (planar sheets); and pipes (cylindrical bodies). Where the contacts between an intrusion and the intruded wall rocks, or country rocks, are parallel to layering in the latter, the igneous body is termed concordant. In contrast, the marginal contacts of a discordant intrusion truncate the layering in the pre-existing wall rocks.

Subjacent bodies are at least partly discordant, and many have steeply outward-dipping contacts. The cross sections of many such plutons (intrusives of irregular or unknown shape) show an increase in width at greater crustal depth. Subjacent masses are termed batholiths if their present outcrop areas exceed about 300 square kilometers, stocks or cupolas if less extensive. The terminology for a specific body is of course dependent on the present erosional level. The great majority of subjacent bodies consist dominantly of intermediate and silicic rock types. Felsic batholiths are exposed in the cores of many eroded mountain systems around the world, where they are confined to the thickened crust of continents and island arcs. North American examples include the Sierra Nevada and Peninsular Range batholiths of California and Baja, Mexico, the Boulder and Idaho batholiths of Montana and Idaho, and the Coast Range batholith of British Columbia. A diagrammatic cross section of a long-since-congealed, and partly eroded, plutonic complex is illustrated in figure 4.4. Roof pendants and inclusions are bodies of country rock partly or completely surrounded by the pluton. While the upper crustal portions of batholiths enlarge downward, they must terminate within the lower crust, hence their overall cross-sections (never well-exposed) must be roughly mushroom shaped.

Tabular intrusions are of two distinct varieties, discordant dikes and concordant sills. Dikes appear to have been intruded along fissures cutting across the pre-existing layering of the host rocks. Dike walls are nearly parallel, and their lengths far exceed their widths (see figure 4.5A). Some dikes are very thin, having widths on the order of a centimeter; others are tens of meters thick. The world's largest known dike, the Great Dyke of Rhodesia (Zimbabwe), is a whopper approximately 500 kilometers in length, and ranges from three to twelve kilometers in width. All major magma types occur in dikes.

Sills appear to have been intruded along fractures coincident with

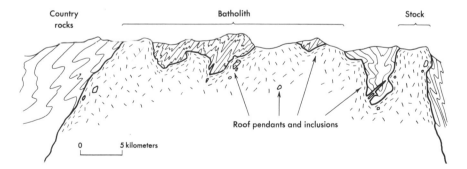

FIGURE 4.4. Diagrammatic cross section through a large subjacent, downward-enlarging pluton such as the Sierra Nevada batholith (Ernst 1969). Roof pendants are large inclusions of invaded wall rocks (country rocks) that hang down into the igneous body.

the pre-existing layering of the country rock (see figure 4.5B). As in dikes, examples of all principal magma types occur in sills. They are also similar to dikes in their proportions of length to width. Dikes and sills are often associated, as in the Hebrides, South Africa, Antarctica, and the Triassic faulted basins of U.S. Atlantic coastal states. Special varieties of sills, illustrated in figure 4.5C and D, include (1) laccoliths, small intrusions with a domical upper contact due to arching of the overlying country rocks and a nearly planar floor and (2) lopoliths, enormous floored intrusions, exhibiting curved concordant contacts typically concave upwards. Laccoliths generally consist of the more silicic, felsic rock types, as in the Henry Mountains, Utah. On the other hand, lopoliths

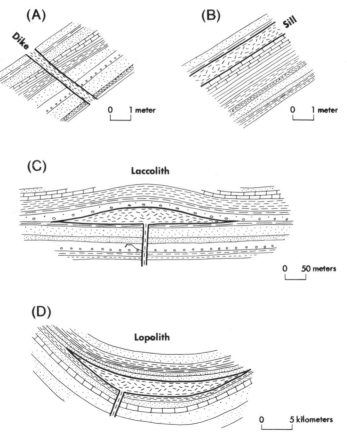

FIGURE 4.5. Schematic cross sections of various tabular and lenslike intrusives; for simplicity, only single-stage, as opposed to multiple-stage, composite intrusions, are shown (Ernst 1969). Note differences in scale.

result from the intrusion of predominantly mafic magma, as very roughly approximated by the Duluth complex, Minnesota.

The third type of intrusive body to be described includes relatively small, discordant plutons only. The chief representatives of this group are cylindrical intrusions known as *pipes*, commonly less than a kilometer across. Many are probably the filled conduits of eroded volcanoes, or volcanic necks. As such, they may consist of any rock type that occurs as lava. Volcanic necks commonly are associated with dikes that radiate out from the central throat, as illustrated in figure 4.6. Certain other distinctly different plutons represent cross cutting, roughly teardrop-shaped masses of rock that have worked their way upward toward the surface in a largely solid or plastic condition. To this category belong a variety of ultramafic rocks (that is, rocks that contain more dark minerals and less SiO_2 than the mafic, subsilicic types) including partially hydrated peridotites (kimberlites) referred to briefly in chapter 3 in connection with the origin of diamonds.

It should be added that nearly all of the larger intrusive masses are composite: that is, each represents the emplacement of more than one

FIGURE 4.6. Shiprock, Navajo Nation, New Mexico, an eroded volcanic neck; note radial, nearly vertical dikes. (Courtesy of the American Museum of Natural History.)

body of magma. Batholiths are especially complex sequences of rock types, and it would be a gross oversimplification to consider the igneous history of any such mass as involving a single injection of consanguineous, homogeneous melt.

The intrusive rocks are coarser grained than chemically equivalent lavas. Intergradations exist, but average grain sizes of the former generally exceed five milimeters, whereas the latter consist of interlocking crystals commonly one milimeter or smaller in dimension. The ultimate size of a specific mineral grain depends on the amount of material that reaches its growing surfaces and is deposited during crystallization. This in turn is a function of the concentration of the various components in the molten solution, the ease with which fresh melt circulates to the environment of the growing crystal, the diffusion velocities of the needed components through the magma, and the rate of cooling. High temperature, a low cooling rate, and the presence of volatile constituents that increase fluidity, circulation, and diffusion all promote large grain size.

Although lavas are extruded at temperatures approaching or exceeding 1000 degrees Celsius, during depressurization on ascent their volatiles are given off abruptly—even explosively—and the melt is rapidly cooled, or quenched; thus most lavas are quite fine-grained. On the other hand, deep-seated, intrusive magmas cool very slowly, (on the order of a few degrees Celcius per million years), and, through retention of volatiles, they remain fluid to lower temperatures. (The presence of dissolved volatile constituents lowers the temperature at which a melt freezes, as shown in figure 7.3.) The cooling episode is much longer, thereby accounting for the larger average grain size of plutonic rocks compared to volcanics. The margin of a pluton is the only part that can cool rapidly, and relatively finer-grained chilled contacts may therefore develop adjacent to the lower-temperature country rocks. Especially among silicic intrusions, the last melt to freeze may be highly charged with whatever volatiles are retained from the initial melt. This material, a pressurized but tenuous aqueous fluid, sometimes crystallizes to an extremely coarse-grained quartz and alkali feldspar-rich aggregate known as pegmatite. Many gem minerals, such as tourmaline, are quarried from such coarse-grained veins and pockets.

Chemical and Mineralogic Classification

Igneous rocks are quantitatively distinguished based on their contrasting chemical compositions. Because the bulk-rock chemistry is reflected in the nature and proportions of the phases present, we will utilize a very simple classification based on the proportions of quartz, feldspars, and dark, ferromagnesian minerals, and on the composition of

TABLE 4.1. Mineralogic Characteristics of Major Igneous Rock Types

Rock Name		Mafic Minerals	Approximate Percentages of Minerals				Composition of Plagioclase
intrusive	extrusive		mafics	quartz	K-feldspar	plagioclase	Plagioclase
granite	rhyolite	biotite ± hornblende	10	35	25	30	sodic
grano-diorite	dacite	hornblende ± biotite ± clino-pyroxene	25	20	15	40	inter-mediate
diorite	andesite	hornblende ± ortho-pyroxene ± clino-pyroxene	35	5	10	50	inter-mediate
gabbro*	basalt	clino-pyroxene ± ortho-pyroxene ± olivine	45	0	5	50	calcic
perido-tite	komatiite	olivine ± ortho-pyroxene ± clino-pyroxene	80	0	0	20	calcic

* Shallow intrusive dikes and sills are termed diabase

the plagioclase. The rock types are briefly described in table 4.1. Although percentages of minerals are listed, it must be remembered that a nearly continuous spectrum of compositions is reflected in the igneous rock series, and the values presented have wide, somewhat overlapping ranges in phase proportions. A slightly more complicated scheme, based on the proportions of the ternary feldspars, their compositions, and the amount of quartz, is presented in figure 4.7.

Because the classification of igneous rocks depends on the composition of the plagioclase, and the proportions of K-feldspar—both difficult to measure without a microscope, X ray machine, or yet more sophisticated instrument—recognition of rock type in the field is not always an easy task. Fortunately, the general proportion of dark minerals also varies directly with the calcium (or anorthite) content of the plagioclase, and inversely with the quartz content, hence the color index is quite helpful in the macroscopic identification of hand specimens.

Bulk compositions reflect differences in mineral proportions among the various rock types. Compared with mafic intrusives, felsic plutons possess greater amounts of SiO_2, K_2O, and Na_2O, and lesser amounts of CaO, FeO, and MgO, reflecting the abundance of quartz, sodic, and potassic feldspar and the general scarcity of ferromagnesian minerals. Chemical analyses of lavas reveal compositional ranges corresponding to those of intrusive rocks. Average, approximate compositions of the oceanic crust, the continental crust, and several important sedimentary rocks are presented for reference in table 4.2.

Certain coarse-grained plutonic rocks, however, have no chemical equivalents among the extrusives. Typical examples are dunite, most peridotite, pyroxenite, and anorthosite, which consist of olivine, olivine and

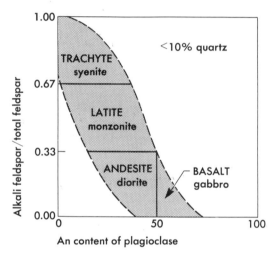

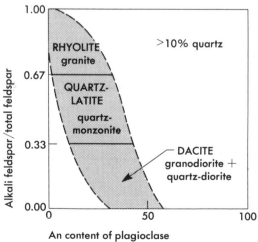

FIGURE 4.7. Classification of igneous rocks. Extrusives are denoted by UPPER-CASE, intrusives by lower-case, letters. Heavy lines indicate classification boundaries; patterned areas show approximate ranges of common rock types. A continuous spectrum of compositions exists, rendering any such classification scheme somewhat arbitrary. An stands for anorthite content of plagioclase feldspar.

TABLE 4.2. Comparison of the Chemical Compositions in Weight Percentages of the Average Oceanic and Continental Igneous Rocks and Average Sedimentary Rocks of the Continental Crust

Oxide	Average Ocean-floor Basalt[a]	Average Continental Igneous Rock[b]	Average Sandstone[b]	Average Shale[b]	Average Limestone[b]
SiO_2	49.56	59.43	78.42	58.46	5.20
TiO_2	1.42	1.05	0.25	0.66	0.06
Al_2O_3	16.09	15.42	4.78	15.51	0.81
Fe_2O_3	—	3.09	1.07	4.05	0.54
FeO	10.17	3.82	0.30	2.47	0.00
MgO	7.69	3.51	1.16	2.46	7.89
CaO	11.34	5.11	5.51	3.14	42.59
Na_2O	2.80	3.86	0.45	1.31	0.05
K_2O	0.24	3.15	1.31	3.27	0.33
H_2O^+	—	1.16	1.63	5.04	0.77
P_2O_5	—	0.30	0.08	0.17	0.04
CO_2	—	0.10	5.04	2.65	41.56
C	—	0.00	0.00	0.81	0.00

[a] Pearce 1976; total iron cast as FeO.
[b] Clark 1924.

pyroxene(s), pyroxene(s), and plagioclase, respectively. Some of these rocks represent layers of accumulated crystals within a magma chamber in the crust; others evidently originated in the mantle and subsequently have been emplaced in the crust in a semisolid state. In either case such rocks, which commonly are monomineralic or bimineralic, do not crystallize under crustal conditions from magmas of their own bulk compositions, as is evident from the fact that lavas of corresponding chemistry have never been found. The sole exception appears to be the so-called komatiites: ultramafic, mantlelike lavas that are virtually confined to ancient, primitive portions of the crust (see chapter 7), and probably represent large degrees of peridotite partial melting in the early, hotter Earth.

But how do differences in bulk chemistries of igneous rock series come about? Fundamentally, the phenomenon of generating lithologies of differing compositions is a consequence of the unequal partitioning, or igneous fractionation (differential separation) of elements between melt and crystalline phases, followed by the spatial removal of liquid from the parental assemblage of minerals. This can happen during heating and partial fusion of a protolith, for instance. Once a melt is generated, it may leave behind the relatively dense, refractory crystalline residuum and ascend into the crust, where it will lose heat. On cooling, solid accumulations of new, early-formed crystals tend to be initially enriched in refractory constituents, whereas the late, low-temperature magmas are charged with the more fusible components. The process is known as igneous differentiation, and explains the contrasts in compositions of melts and solids produced during either partial fusion or partial freezing—or more commonly, both.

Plate-tectonic Settings

First, we now examine the manner in which igneous rocks are formed in the ocean basins; then we will consider continental magmatism. The production of melt in the oceanic upper mantle appears to be relatively straightforward, and therefore is reasonably well understood. In contrast, the mechanisms for generation of magmas that characterize the sialic crust are far more complex, and thus are correspondingly more controversial and obscure.

The ocean basins are floored predominantly by basaltic crust, and its intrusive equivalent, gabbro. These have been produced in the vicinity of a midoceanic ridge, the near-surface manifestation of rising mantle currents. Why does partial melting occur within the ascending limb of a convection cell? For the answer, we must turn to the study of mineralogic phase equilibrium.

A pressure-temperature diagram for the anhydrous (waterless) partial melting of homogeneous peridotite approaching chemically undifferentiated, or primitive, mantle, and similar to stony meteorites in composition, is shown as figure 4.8. This figure reproduces the very lowest pressure part of the thermal-depth regime for the entire Earth depicted in figure 1.10. The onset of melting of such material climbs to higher temperatures as pressure (depth) increases. This is a consequence of the fact that, in general, melt has a larger volume than the solid from which it is derived, and increasing pressure favors the small-volume, solid assemblage. On the other hand, as plastic, ductile mantle material near, but at temperatures slightly below, the melting curve rises towards the surface, pressure decreases; however, because rocks are good thermal insulators (that is, they are poor conductors of heat), the elevated temperatures remain almost constant. Thus, the ascending mantle beneath a spreading ridge will eventually intersect the partial fusion curve, and incipient melting will ensue. This process is called decompression melting, inasmuch as it occurs in response to a pressure drop at nearly constant temperature. Once formed, melt tends to separate from the slowly rising mantle sheet (or convective cell limb) because the derived liquid is less dense than the now depleted, largely solid, parental peridotite. Basaltic magma probably forms above a local asthenospheric hot spot in response to the same decompression of a rising, cylindrical mantle plume at nearly constant, high impact and temperature. (Ascension may result from heating from below or meteorite excavation from above.)

The chemistry of the magma produced from relatively undepleted mantle material varies as a function of the pressure of formation and the degree of partial melting, which in turn is proportional to the temperature above the beginning of melting curve. Low-pressure (zero- to ten-kilobar) melts tend to have compositions similar to those of common oceanic basalts and gabbros (see table 4.1). Higher-pressure liquids are impoverished in silica relative to normal oceanic basalts, and with low extents of incipient fusion tend to possess rather alkalic chemistries; and at high temperatures and pressures (30 to 50 kilobar) and/or at substantial degrees of partial fusion in the mantle, magmas are especially olivine-rich (i.e., they are peridotitic, or komatiitic).

The separated mafic magma rises towards the rifted sea floor in and near the axis of the spreading center. Material that reaches the sea bottom is extruded as massive basalt flows, pillow lavas, and fragmented flows (breccias). That which crystallizes at depth forms subjacent gabbroic units; crystal settling of denser refractory minerals results in the accumulation of mafic and ultramafic layers—termed cumulates —at the base of the magma chamber. Continued supply of melt to the rifting, fracturing sequence of oceanic crust results in the injection of

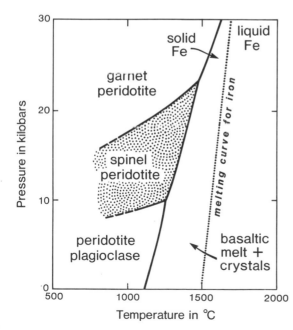

FIGURE 4.8. Anhydrous, solid-state, and solid-liquid equilibria in relatively primitive mantle material approaching chondritic meteorites in chemistry. Major phases in periodotite are olivine-plus-pyroxenes. Calcic plagioclase, or spinel ($MgAl_2O_4$), or magnesian garnet are minor aluminum-bearing phases indicative of particular pressure-temperature stability regions. Incipient melting produces a basaltic magma, especially at relatively low pressures. For reference, the melting curve for pure iron, indicated by dotted lines, is projected into the diagram.

diabasic dikes and sills into the ridge superstructure, where they cool more rapidly than the subjacent gabbros, but more slowly than the overlying basaltic lavas. Diabases are therefore rocks texturally and structurally intermediate between basalts and gabbros. As the oceanic crust moves away from the ridge crest, settling of fine-grained debris from the overlying column of seawater provides a thin capping layer of deep-sea sediments, which gradually accumulates with the passage of time. Generalized relationships, illustrated in figure 2.6, are shown along with the presumed thermal structure of a divergent plate boundary in figure 4.9. Where relatively intact, complete sections of oceanic crust plus uppermost mantle have been thrust into island-arc or continental-margin crustal environments, they are called ophiolites.

The formation of oceanic crust is a reflection of the mantle's attempt

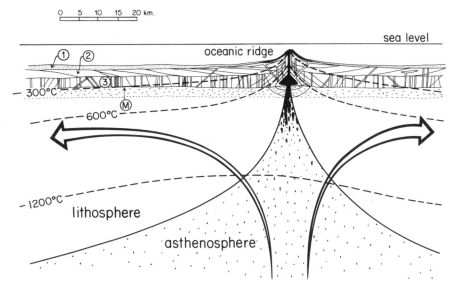

FIGURE 4.9. Schematic diagram of a divergent plate margin or spreading center, after figure 2.6. The thermal structure is also depicted. Heat flow is high at and near the oceanic ridge, and gradually lessens laterally away from the active spreading center. Increasing degrees of anhydrous partial melting of the asthenosphere during ascent and depressurization provides basaltic midocean ridge magmas shown in black. Oceanic layers are: (1) deep-sea sediments; (2) pillow basalts and breccias; and (3) gabbro-plus-diabase complexes. (M) is the Mohorovicic discontinuity. The basal region of cumulate ultramafic material (random dash pattern) occurs as an early crystallization/accumulation product of the magma along the floor of the magma chamber. Large arrows indicate inferred directions of mantle flow and lithospheric plate transport.

to rid itself of excess heat. The phenomenon involves a gravitative instability consisting of a dense lithosphere surmounting less dense asthenosphere, coupled with decompression-induced partial fusion of the buoyant, rising, plastic mantle plume or sheet. Whether the circulation is
driven by a deep-seated thermal instability in the mantle or by the near-
surface plates sliding off the ridges, or being pulled down the trenches is
still debated. Probably all these mechanisms are involved to a greater or
lesser degree in the movement of the plates and the mantle flow regime.

But how do we account for the primary igneous processes that have
produced the continental crust over the course of geologic time? (For
estimates of the average bulk composition of continental igneous rocks,
see table 4.2.) First, let us observe that the more silicic active volcanoes
and relatively young, subjacent batholiths are disposed around the margin of the Pacific, and also discontinuously mark the eastern Mediterra-
nean-Indonesian mountain chain. Everywhere today, these high-heat
flow regimes are located on stable, nonsubducted lithospheric plates, set
back 100–200 kilometers from the deep-sea trenches, the surface expression of convergent plate boundaries. Such volcanic belts are situated
approximately 100 kilometers vertically above the descending lithospheric slabs; individual igneous centers are sited at roughly periodic
intervals of 50–100 kilometers along the longitudinal extent of the volcanic arc. The partial fusion of oceanic crust capping the downgoing
plate, or incipient melting of the overlying peridotite of the stable, nonsubducted lithosphere, or both, followed by upward buoyant migration
of molten rock, are thought to account for construction of the magmatic
belts. The characteristic liquids so formed beneath modern island arcs
and continental margins are of andesitic to dacitic chemistry, but they
are accompanied by basaltic melts as well. Perhaps the basalts represent
the small amounts of molten materials that wet grain-boundary surfaces
throughout the asthenosphere and have been drained upwards from
directly beneath the stable, nonsubducted lithospheric plate, whereas
the andesites-dacites are derived from the partial fusion of subducted
oceanic crust capping the downgoing slab.

If this model is correct, then andesites and more silicic liquids characteristic of the sialic crust require two stages of planetary partial melting for their shallow generation: first from upwelling, decompressing
mantle peridotite at ten- to thirty-kilometer depths or more in the axial
zone of an oceanic spreading center to produce sima; then from deeply
subducting, compressing oceanic basalt about a hundred kilometers down
along an inclined, convergent plate junction. This scenario would explain the ultimate, gradual production and accretion of continental crust
from a thermally and/or gravitatively convecting, chemically differen-
tiating mantle through a two-stage process. Of course, although we have

spoken of the magmas as if they were all extrusive members of a particular chemical and mineralogic type, some such liquids crystallize at depth to form the corresponding subjacent plutonic series of igneous rocks.

Unlike the basaltic oceanic crust, which is almost completely reintroduced back into the mantle during continued mantle circulation, the lower-density, hence strongly buoyant, continental crust appears to be nearly unsubductable. Accordingly, it has gradually accumulated over the course of geologic time, thereby accounting for the disparity in average ages of the continents and ocean basins mentioned in chapter 2. With deep burial, long annealing (slow cooling and recrystallization after subjection to heat) times, and repeated metamorphism/deformation, deeper portions of sialic mountain roots evolve toward the refractory, anhydrous granitic gneiss associations characteristic of ancient continental shield areas, as now exhumed and exposed on the stable cratons, whereas the fusible, volatile-rich, molten fractions are displaced upward into middle and upper levels of the continents to form the batholithic complexes.

Relationships for a convergent plate junction, illustrated diagrammatically in figure 2.7, are depicted along with the inferred thermal structure in figure 4.10. Also shown in this figure is the partial fusion of thickened, basal portions of the more "juicy" continental crust. The presence of abundant aqueous fluids in such preexisting rocks (sodium-, potassium-, and silicon-rich sedimentary rocks, siliceous igneous units and the metamorphosed equivalents of both) promotes partial melting and upward buoyant ascent at moderate temperatures and pressures— say, 650 degrees Celcius and ten kilobars—because the solubility of H_2O in the granitic magma lowers its temperature of melting several hundred degrees Celsius (see figure 7.3). Therefore, evolving island arcs and continental crust situated at convergent plate junctions contain both primary igneous materials derived from the mantle through two different stages of partial melting, and secondary (crustally remelted) magmas formed through multicycle sedimentary, metamorphic, and igneous processes.

⎯ SEDIMENTARY ROCKS

Occurrence of Sedimentary Rocks

Most estimates of the crustal volumes for the major rock types include on the order of 5–10 percent sedimentary strata. These units occur as a nearly ubiquitous, but generally thin surficial covering of both continents and ocean floors. Locally, along rifted continental margins and near convergent plate junctions, the thickness of sedimentary basinal deposits may approach or even exceed fifteen kilometers. Sedimentary units are of great economic importance, for in them is sequestered much of the world's readily extractable mineral wealth, including coal, petroleum, natural gas, nuclear fuels, aluminum, iron, and manganese ores, and building materials such as stone, sand, gravel, marl, and limestone (see chapter 6). Indeed, many metal deposits of, for instance, copper, zinc, and lead, which appear to be genetically related to magmas,

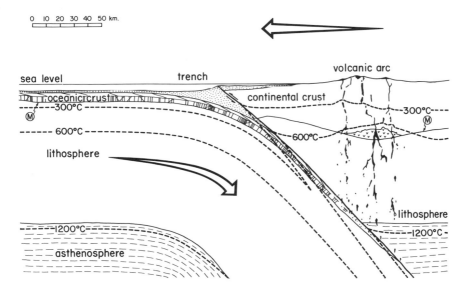

FIGURE 4.10. Schematic diagram of a convergent plate margin, after figure 2.7. The thermal structure is also illustrated. Heat flow is low in the vicinity of the trench, and gradually rises to high values over the broad region of the volcanic (plus plutonic) arc. (M) is the Mohorovicic discontinuity. Partial melting of pre-existing rocks is inferred to occur in three distinct regions: (1) the recrystallized basaltic (plus gabbroic) oceanic crust at the top of the downgoing plate; (2) the upper part of the stable, nonsubducted asthenosphere through which volatiles are presumed to be migrating above the adjacent descending plate (the source of the fluids); and (3) the basal, thickened portions of the sialic, H_2O-rich crust. The first two primary magma types are shown in black, and include both basaltic and andesitic-dacitic compositions; the third type of magma, indicated by a checked pattern, represents a relatively granitic, secondary plutonic series produced by partial fusion of deeply buried old continental rocks in the presence of volatiles. The width of the zone of active igneous activity in the arc at any one time is somewhat less than the composite field shown. Large arrows indicate major directional components of lithospheric plate motion.

are localized at the contacts between igneous intrusions and chemically dissimilar metasedimentary rocks (i.e., recrystallized sedimentary units).

Sediments are accumulations of materials that have been reworked by surficial processes from pre-existing rocks of any origin. Geologists broadly distinguish between (1) clastic sediments, or detritus—the products of mechanical accumulation of individual grains and (2) chemical sediments—materials that have precipitated from inorganic solutions. Bioclastic and biochemical deposits are analogous sediments produced largely through biological interactions. Of course, most clastic rocks contain some chemically precipitated material, and most chemical sediments carry clastic fragments. All sedimentary rocks result from the following processes: weathering of source materials; transportation, commonly in water; deposition by organic or inorganic means; and lithification, that is, the process of natural compaction and cementation whereby the original unconsolidated sediment is converted into a more coherent, cohesive rock.

The nature of a sediment will depend in part on that of the source area from which it was derived, and on the attending weathering processes. Weathering is a form of surface alteration, and involves the mechanical and chemical breakdown of a protolith (the original rock prior to alteration), usually by interaction with the atmosphere, or with water. When pre-existing rocks are weathered, they commonly form a soil profile in which three intergradational horizons or layers, generally referred to as A, B, and C horizons, can be recognized. The surficial zone A is enriched in organic debris but is leached of alkalis, other soluble cations such as magnesium and in some cases calcium, and the soluble anion complexes of carbon, phosphorus, nitrogen, and sulfur. This dissolution occurs as a result of the downward percolation of rain water, which contains organic acids but is not yet saturated with respect to soluble constituents. In contrast, the intermediate zone B is enriched in clay minerals and hydrated iron oxide because of the precipitation and accumulation of dissolved material from the downward percolating aqueous solutions which, at this stage, are saturated. Relatively fresh but fractured, disaggregated protolith constitutes the lowest zone C.

The depth to which a soil profile develops is a function of the rates of chemical weathering plus solution, and mechanical breakdown plus bodily removal. These competing rates in turn depend on factors such as topography, climate, and organic activity. High topographic relief favors rapid mechanical disaggregation and erosion, whereas low relief inhibits it. Chemical reactions are accelerated by high temperatures and most require the presence of an aqueous fluid, so elevated temperatures, high rainfall and/or low evaporation rates promote chemical weathering. Such climates promote luxuriant plant and animal life, and the rapid decay of organic matter. Cold, rigorous climates inhibit chemical activity because of low temperatures, the lack of mobile aqueous solutions, and an impoverished flora and fauna; however, frost action and glaciation are effective erosive agents, through fracturing, spallation (breaking off in chips, scales, and slabs), abrasion, and bodily movement during freezing and thawing.

The state of aggregation of a sediment depends to a considerable extent on the mode of transportation from the source to the depositional area. Mass movement represents a response to the force of gravity and includes the accumulation of talus—large, angular fragments of rock derived from disintegrating cliffs and steep slopes. Mudflows are another variety of mass movement, one that results, however, in the aggregation of much finer-sized materials. Avalanches and landslides have properties more or less intermediate to these two extremes. Particles entrained in a transporting medium such as air, water, or ice are abraded as they rub against one another and are moved over the underlying material; accordingly grain sizes and angularities are diminished as a function of the distance carried. The dissolved products of chemical attack are moved from the parent rocks to the site of deposition exclusively in aqueous solution. Running water is also the prime carrier of fragmental debris.

The manner of deposition exerts a profound influence over the aspect of the resultant sediment. Rapid accumulation of ill-sorted clastic (detrital) debris of several size ranges generally results in a weakly laminated or massive deposit; on the other hand, slow buildup of particulate matter, or of chemical precipitation alternating with the settling of fine clastic debris, tends to produce finely bedded sediments. The rate of discharge of material from the transporting medium is a function of the flow velocity and nature of the transported sediment: both in turn are related to topography, climate, vegetation, and the mineralogic and chemical constitution of the source, as discussed above.

Two principal types of sediment are distinguished, based on the site of deposition, (1) continental, or terrestrial, and (2) marine. Among the former are the products of subaerial mass movements, including talus (clastic debris accumulated at the base of a cliff or slope), landslides and avalanche chaos, stream gravels, lake beds, sand dunes, and glacial deposits. Red beds are clastic rocks in which some of the iron in the constituent minerals has been oxidized, either during or after lithification (see next paragraph); many but not all are deposited in a continental setting. All such sediments, with the possible exception of some lake beds, are composed mainly of clastic units that grade laterally and vertically rather abruptly into other sedimentary rock types, reflecting rapidly fluctuating depositional conditions on land. Marine sediments, on the other hand, tend to be more widespread and continuous, reflect-

ing the more gradually varying subsea environment. To this group belong shallow-water deposits of deltas, beaches, continental shelves, and
slopes, as well as deep-sea sediments of the ocean basins. Both chemical
precipitates and clastic materials are abundant in all such sediments.
Four main groups of broadly disposed and areally extensive marine
strata may be recognized. The first is the platform group, the term
relating to certain continental shelves, gulfs, and banks; their bodies of
water are known as epeiric seas. The strata in question are characterized
by thin units of shallow-water origin. The second is the miogeoclinal
(continental slope and rise) group, having thin to thick, well-ordered
deeper-water deposits. The third is the eugeoclinal group, which typically occurs as thick, in some cases chaotic mixtures; the group consists
of relatively deep-water deposits containing volcanics (generally associated with oceanic trench environments). The final group is called hemipelagic (deep sea) and is marked by very thin, finely laminated clays and
chemical precipitates.

The process of lithification, whereby an unconsolidated sediment is
converted into a coherent rock, involves several distinct but intergradational phenomena. Compaction of a sediment due to the weight of the
overlying rocks (overburden) decreases pore space and results in the
deformation of soft mineral grains, clast rotation, flattening, squeezing
out of interstitial water, pressure solution, and in the production of some
interlocking of grain boundaries. Iron oxide, silica, and/or calcium carbonate minerals precipitate from pore solutions and act as cement to
bind the particles together. Reaction among the original, chemically
dissimilar fragments produces new minerals that transect former grain
boundaries and increase rock coherence. This recrystallization is known
as diagenesis when it takes place in the original basin of deposition,
usually directly following sedimentation. As the original sedimentary
basin is gradually obliterated, diagenetic transformation gives way imperceptibly to metamorphic recrystallization. This latter process is
thought to take place generally at somewhat higher temperatures and
pressures (in other words, more deeply buried in the Earth) than diagenesis, although the two processes are completely intergradational.

Textural Classification

In order to discuss the chemical and mineralogic variations of the
different stratified rocks, we need to adopt a systematic nomenclature.
Sediments are classified on the basis of their grain sizes, chemical and
mineralogic compositions, and modes of accumulation. Classification by
grain size is most readily applied to clastic sediments because particle
size is related to the origin of the rock. Variations in the bulk composi-

tions and mineralogies of clastic sediments are also related to grain size, reflecting contrasts in chemical and mechanical stabilities of different minerals. In chemically precipitated deposits, grain size is less useful for classification because postdepositional recrystallization frequently causes a coarsening of the particles. Grain sizes are not readily correlated, therefore, with compositional ranges among chemically precipitated sediments.

Among the clastic rocks, we define four main grain-size groups: (1) conglomerates and gravels containing numerous particles greater than two millimeters in diameter (where the individual fragments are angular, the rocks are called sedimentary breccias); (2) the 1/16- to two-millimeter size range, where the most abundant grains of sandstones fall; (3) siltstones containing grains between 1/256 and 1/16 millimeter in diameter; and (4) claystones carrying particles finer than 1/256 millimeter in diameter. Where parting, or fracturing, along bedding surfaces, (fissility) is developed, claystones and fine-grained siltstones collectively are termed shale. If fissility is not present in a deposit containing almost equal proportions of grains in both clay and silt size ranges, the somewhat massive rock is termed mudstone. Commonly a sediment contains grains of several size ranges, in which case the deposit is named on the basis of the dominant (most abundant) particle size.

For clastic carbonate rocks, the terms calcirudite, calcarenite, and calcilutite signify size ranges corresponding respectively to those of conglomerates plus gravels, sandstones, and shales (siltstones and claystones). Limestones produced principally through the accumulation of calcareous fossil materials include coquina (sorted shell debris), chalk (mainly exoskeletons of foraminifera, one-celled marine organisms), and bioherms, or reef limestones. Except for chalk, most fossil fragments display moderately coarse particle sizes. In contrast, the inorganically deposited carbonates and siliceous sedimentary rocks such as lithographic limestone (chemically precipitated $CaCO_3$), chert (chemically precipitated silica), and banded iron formation (bedded iron oxide and silica) are typically very fine-grained, and are named without reference to particle size.

Chemical and Mineralogic Diversity

As is clear from the description presented earlier in this chapter, compositional variations among the igneous rocks are relatively modest, because fractionation of elements between coexisting melt and crystals, although systematic, is not pronounced. Thus igneous crystallization produces a relatively narrow spectrum of related magmatic rocks of somewhat similar chemistries. In contrast to these primary rock types, sediments exhibit a strikingly broad range of bulk compositions, as is illustrated in figure 4.11. The reason for this is that both mechanical and chemical properties of the constituent minerals, as well as erosional and transportational histories, markedly influence the composition of the final sedimentary deposit. Mechanical characteristics of minerals are very different from one another. The process whereby a more or less compositionally uniform source is converted to secondary accumulations of contrasting compositions is termed sedimentary differentiation.

Weathering of pre-existing rocks at or near the Earth's surface results in the solution of readily dissolved constituents and in the hydration and oxidation of others. The exact nature of the residual weathering products depends on factors discussed previously. Materials transported as clastic grains are further separated on the basis of mechanical properties; relatively soft minerals possessing good cleavages are more rapidly and more finely abraded than hard mineral fragments lacking fractures and cleavage. Therefore, grain size and chemical (mineralogic) composition are related in a sediment. Sand grains consist predominantly of quartz, which may or may not be accompanied by feldspar(s), whereas clay-sized particles are mostly aluminous sheet silicates; as a

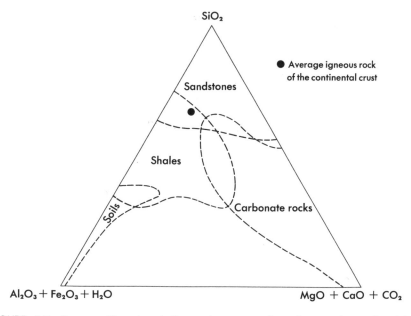

FIGURE 4.11. Compositional variations of some sedimentary rocks and residual soils (After Mason 1966, figure 6.2.)

result, the size sorting that occurs during transport results in a distinct chemical separation as well.

The process of deposition accentuates this chemical differentiation, with coarser-grained feldspathic (feldspar-comprised) and especially, quartz-rich (quartz being very resistant) gravels and sands being laid down in agitated shallow water much nearer to the source than the more finely divided silt and clay particles, which are carried successively further from the parent rocks and ultimately settle out in quiet, deep-water environments. Moreover, near-shore reworking by wave and current action tends to break down all grains but the most stable, both chemically and mechanically; as a result, residual quartz is enriched in such deposits because of its superior hardness, lack of cleavage, and resistance to corrosion. Chemically and biochemically precipitated materials tend to be diluted by the rapid accumulation of coarser-grained clastics in the near-shore regime, but as the particle sizes of fragments diminish in quieter waters, clastic deposition rates decline, and chemically precipitated materials become volumetrically more important, generally at greater distances from the source of particulate debris. Finally, after deposition, the processes of diagenesis (recrystallization) and interstratal solution (dissolution and precipitation by aqueous fluid passing through the bed) may further change the bulk composition of the rock by dissolving unstable particles and precipitating an interstitial cement.

As indicated in table 4.2, the average compositions of igneous source rocks of the continental crust and of typical sedimentary lithologies, sandstone, shale, and limestone, are markedly different, reflecting contrasting mineralogies. Sand-sized particles are enriched in quartz and feldspars, which accounts for the high silica contents of sandstones. In contrast, shales carry abundant clay minerals which are relatively low in silica, but high in alumina, alkalis (especially potash), and H_2O. Limestones are enriched strongly in $CaCO_3$ and $MgCO_3$. However, many clastic rocks carry moderate amounts of carbonate and most limestones contain minor quantities of quartz and clay minerals.

The average bulk composition of Phanerozoic continental sedimentary rock, which is about 60 to 80 percent shale, 15 to 20 percent sandstone and conglomerate, and approximately 10 to 15 percent carbonate, corresponds rather well to the average continental igneous rock, except for volatile constituents. These latter components, especially H_2O and CO_2, undoubtedly were initially present in the primary magmas. They were given off, however, on igneous crystallization, hence are concentrated in the atmosphere, in the seas, and in secondary rocks. However, the aggregate compositions of the sedimentary veneer and of the atmosphere-plus-hydrosphere have evolved with time, as well be discussed briefly in chapter 7.

Multicycle sediments are those that have been deposited, then eroded, transported, and redeposited, in some cases several times. As a consequence, such mature sediments exhibit great chemical differences from one another. First-cycle, immature sediments, in contrast, have bulk compositions not greatly disparate from the source terranes because of short transport distances and limited time for chemical and mechanical attack.

Plate-Tectonic Settings

We have seen that the grain sizes and mineralogic constitutions of sediments are composite functions of the nature of the source area, attendant weathering conditions, modes of transportation, and sites of deposition. Certain kinds of environments are characterized by specific sedimentary lithologies; recognition of this relationship allows a fuller appreciation of the regional geologic setting and plate-tectonic history of the Earth's crust.

Some of the main types of depositional environment are illustrated in figure 4.12. Not shown are the stable cratons (continental interiors), home of the shallow-water epeiric seas and platform strata as well as Precambrian basement; nor are deep-sea environments (ocean bottoms far distant from continents and island arcs), the locus of thinly laminated, fine-grained abyssal sediments, figured. The former realm is the site of deposition of well-sorted, multicycle clastic deposits, in which chemically distinctive quartzose (quartz-comprised) sandstones are associated with aluminous clay-rich shales and platform carbonates. These well-ordered, thin sequences of laterally widespread strata are characteristic of the continental shelves. They are produced through processes involving long-continued weathering, and considerable differential transport, reflecting the episodic reworking of pre-existing clastic debris and chemically precipitated units. In contrast, deep-sea deposits typically contain exclusively very fine-grained iron-, aluminum-, and manganese-rich clays interlayered with chemically precipitated SiO_2 (chert). They typically lack $CaCO_3$ because these very thin blanketing strata are laid down below water depths of about 4,000 meters (the so-called calcite compensation depth) where, due to high pressures and low temperatures, calcium carbonate becomes undersaturated in seawater and therefore dissolves. The ages of deep-sea sediments directly overlying oceanic crust in the ocean basins are illustrated in figure 2.15.

A rifted continental margin (Figure 4.12A) is characterized by a thin veneer of landward shelf sediments that gradually gives way seaward to a thick stack of well-stratified, dominantly clastic miogeoclinal strata in the vicinity of the continental slope and continental rise. Outboard from this complex, a substantial accumulation of debris laid down on the deep-sea bottom shows somewhat more oceanic affinities. The eastern seaboard of the U.S. is a good example of this Atlantic type of sedimentary environment.

Convergent margins (figure 4.12B,C) are characterized by well-ordered forearc-basin deposits, and largely chaotic, voluminous melanges of poorly-sorted, rock-fragment-rich feldspathic sandstones, siltstones, and deep-water shales, associated with faulted slices and lenses of oceanic crust as trench basin deposits. This latter assemblage is described as

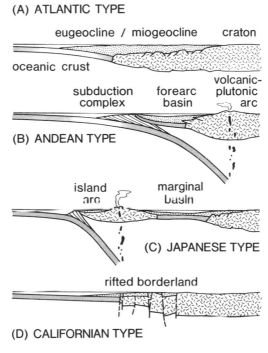

FIGURE 4.12. Various types of continental margins, exhibiting environments of deposition for sediments (Ernst 1979). Former divergent plate margin (A) shows no differential plate motion subsequent to the initial rifting; this passive margin originally was, but is no longer, a plate junction. (B), (C), and (D) are all active margins. Convergent plate margins involve a major component of underflow of the oceanic (rarely continental) crust—capped plate, shown on the left descending beneath the continental margin (B) or island arc (C) on the right. Conservative margin (D) displays differential plate motion obliquely or nearly at right angles to the plane of section.

eugeoclinal, and displays evidence of having been intensely sheared and dragged beneath the stable nonsubducted plate, as is characteristic of subduction zones. The Alaskan-Aleutian arc and Indonesia present modern analogues of this Andean type of depositional setting, reflecting converging plate motions. Marginal basins (figure 4.12C) represent depositional environments in which continental sediments are intermixed with volcanic-arc materials laid down on a basaltic substrate; such series are also referred to as eugeoclinal. The Sea of Japan, situated off the east coast of Asia but behind the active arc, characterizes this Japan-type sedimentary basinal configuration.

Continental transform basins (figure 4.12D) generally consist of an interlocking mosaic of crustal blocks jostled about vertically and horizontally, reflecting differential shear between the lithospheric plates. For this reason, basins of deposition, whether located on sialic or simatic basement, are small, and tend to change geometry rapidly with time and areal extent; consequently, sedimentary deposits in this conservative plate regime laterally pass abruptly from conglomerates and lithic fragment-rich sandstones to siltstones and shales. Erosion and deposition rates are very high, hence slowly accumulating carbonate-rich strata are rare. A modern example of this kind of environment is found in the California Coast Ranges and offshore in the southern California borderland.

⎓ METAMORPHIC ROCKS

Occurrence and Classification

Recrystallized and/or deformed pre-existing lithic units constitute roughly 15 percent of the exposed or near-surface continental crust; to a greater or lesser extent, the oceanic crust has been hydrothermally altered, thus may be considered as at least partly metamorphosed. Such lithologies are products of P-T (pressure-temperature) conditions intermediate between those of igneous and sedimentary environments. Thus, on the one hand, metamorphic processes merge with diagenesis, a sedimentary phenomenon; on the other, very intense crustal metamorphism (ultrametamorphism) leads to partial fusion and the generation of sialic magma. Mechanical properties are of major significance under the conditions of low-temperature metamorphism where chemical reactions take place slowly. In contrast, more rapid recrystallization, which drives toward chemical equilibrium, tends to characterize the conditions of medium- and high-temperature metamorphism; under these conditions rocks are weak and deform plastically.

Two principal metamorphic processes, therefore, may be distinguished, mechanical deformation and chemical recrystallization. The former includes grinding, shearing, and plastic deformation of an initial rock, phenomena that reflect readjustment of the material, or strain, as a consequence of differential pressures or stress. Recrystallization takes place because pre-existing mineral assemblages are displaced from chemical equilibrium by changes in the temperature, pressure, or chemical milieu. Nearly all metamorphic rocks show the combined influence of both mechanical deformation and chemical reaction; they differ principally in the degree of development of these effects. Metamorphic rocks exhibit contrasts reflecting differences in bulk composition as well, either due to original differences in chemistry or resulting from alteration accompanying recrystallization.

Planar features are observed in specimens of intensely strained metamorphic rocks. Fine-grained metashales, or slates, break along planar surfaces which are of two types, fracture cleavage and flow cleavage. In the first variety, the rock is characterized by discrete fracture planes that are separated by finite zones of material having no obvious weakness parallel to the fractures. In such rocks, little grain reorientation toward parallelism has taken place. In the second type, a more pervasive, dimensional, and crystallographic alignment of micas and other platy minerals has been produced throughout the material by the deformation. This results in a layering, or foliation, that makes the rock weak in the cleavage direction at every point within it. The orientation of flaky minerals is illustrated diagrammatically in figure 4.13A. Highly aluminous rocks, such as metashales, carry large amounts of micas and chlorites, and display well-developed foliation as a consequence of deformation. Where prismatic and needle-shaped chain silicates such as crystals of amphibole occur in a metamorphic rock, they commonly are oriented dimensionally to produce an alignment, or lineation (figure 4.13B). If foliation is also developed, the lineation direction customarily lies within the plane of foliation. Subsilicic igneous rocks such as basalts have bulk compositions appropriate for the crystallization of large amounts of amphiboles, hence lineation is most marked in the metamorphic equivalents of these rock types. Other linear features include fold axes, corrugations, striations, and intersections of various planar features (such as the intersection of compositional layering with foliation).

In the more intensely recrystallized foliated rocks, an increase in grain size enables us to recognize individual mineral flakes, and a faint tendency for compositional differences to develop in layers parallel to the foliation. Such rocks are termed schists, and the foliation is called schistosity. The coarsest-grained metamorphic rocks, called gneisses, display distinct mineralogic and compositional banding. They usually contain a smaller proportion of oriented, hydroxyl-bearing platy grains,

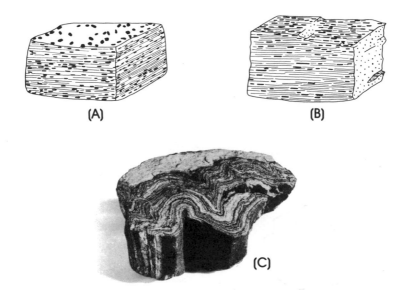

FIGURE 4.13. Sketches of rocks showing (A) platy mineral orientation or foliation, and (B) prismatic mineral orientation or lineation (Ernst 1969). Photo (C) of small-scale folds, or lineations, in granitic gneiss (UCLA collection).

because gneisses form at high temperatures and are rich in anhydrous, refractory phases such as quartz, pyroxenes, and feldspars. Their foliation is termed gneissosity. A contorted gneiss is illustrated in figure 4.13C.

Orientations of linear and planar structures in metamorphic rocks appear to be related to bodily flow and grain rotation accompanying compression and shearing. Furthermore, a stress field may favor the growth of grains with certain crystallographic orientations while disfavoring others. As described in chapter 3, the relative dimensions of prismatic and platy mineral grains reflect their contrasting crystal structures. Therefore, preferred crystallographic orientation produced by recrystallization under stress generally results in the development of dimensional orientations of the mineral grains.

The process of mechanical deformation is called cataclasis, and metamorphic rocks in which comminution (reduction to small particles) is conspicuous are known as cataclastic rocks. Because recrystallization and chemical interaction are not pronounced, planar and linear features in cataclastic rocks must have formed principally through the milling down, shearing, and dislocation of pre-existing materials. The original grains show the effects of plastic strain and severe granulation. Several different types of cataclastic rocks may be distinguished by grain size and degree of recrystallization. Mylonites are thoroughly sheared and milled-down lithologies with grain sizes in approximately the 0.01- to 0.1-millimeter range. Although some mylonites show very little evidence of chemical reaction, others, such as those of the Moine thrust zone of Scotland, exhibit pronounced effects of chemical reaction. Even in recrystallized mylonites, however, shearing outlasts the process of mineral growth, which itself would tend to obliterate cataclastic textures. In general, foliations and lineations are well-developed in mylonitic rocks. Fault gouge is extremely fine clayey cataclastic material lining a zone of differential shear, whereas fault breccia consists of a mixture of coarse, angular, blocky rock fragments as well as milled-down materials.

In contrast to cataclastic rocks, which are produced by dominantly mechanical deformation, contact metamorphic rocks are formed by a pronounced increase in temperature in the virtual absence of differential stress. Contact metamorphic rocks are localized as concentric zones, or aureoles, surrounding hot igneous bodies emplaced at upper levels of the crust. In such environments, the difference in original temperature between what are called country, or wall, rocks and the intrusive magma that invades them is substantial: shallow plutons are intruded at temperatures approaching 1000 degrees Celsius whereas the wall rocks a few kilometers below the surface are relatively cool, having initial temperatures prior to pluton emplacement on the order of 100–200 degrees Celsius. Therefore, a marked thermal gradient exists adjacent to the intrusive contact. Deeper in the crust, the temperature contrast can hardly be so great. We mentioned previously that H_2O-bearing silicic magma may be generated near the base of the continental crust at temperatures approximating 650 degrees Celsius; in the presence of a normal intraplate geothermal gradient, deep-seated country rocks also are relatively warm, with temperatures exceeding 500 degrees Celsius at thirty to forty kilometers depth. Deep levels of the continental crust thus lack pronounced horizontal and vertical thermal gradients and are more commonly characterized by broad zones of regional metamorphism (see next section).

The recrystallization of country rocks adjacent to a shallow pluton does not generally produce preferred mineral orientations because of the absence of penetrative, pervasive deformation. Instead, the grains, many of which are roughly the same size, form an interlocking, nondirectional mosaic. This texture is known as hornfelsic, and the wall rock as a hornfels. Most such contact metamorphic rocks are fine-grained because heating by the magma body does not last long enough for extensive grain growth. Juxtaposed against the igneous contact, however, where the hornfels is heated to the highest temperatures, it may display medium or even coarse grain size. The zonal sequence involves devolatization and a variety of mineral reactions as well as increasing grain size toward the thermal anomaly.

Thus far we have contrasted the dominantly mechanical deformation of cataclastic rocks with the mainly chemical recrystallization of contact metamorphics. The most widespread varieties of metamorphic rocks, however, develop on a regional scale in response to both deformation and mineral reaction. Such widespread rocks are known collectively as dynamothermal or regional metamorphic rocks, and are typified by penetrative, oriented mineral fabrics. Foliations and lineations are widespread features reflecting differential motions (shearing) of the rock sections undergoing recrystallization. Similar to contact metamorphic rocks, regional metamorphics show progressive zonal sequences ranging from lower-temperature, fine-grained, volatile-rich mineral assemblages to higher-temperature, coarse-grained, volatile-poor phase associations. In contrast to contact aureoles, however, the zones developed under regional metamorphic conditions are quite broad, indicating gradual lateral and vertical changes in the physical conditions attending regional recrystallization.

The systematic distribution of characteristic minerals in regionally metamorphosed shales was first documented in the Scottish Highlands by George Barrow. He delineated a series of metamorphic zones, each of which is typified by the occurrence of a critical, or index, mineral. The most feebly recrystallized metamorphic rocks retain clastic micas. At successively higher grades (under more intense conditions of recrystallization), progressive metamorphism results first in the regional development of chlorite (a magnesium-, iron- and aluminum-bearing hydrous layer-silicate) followed by biotite, next almandine garnet ($Fe_3^{+2}Al_2Si_3O_{12}$), then staurolite (a hydrous iron-aluminum silicate), kyanite, and, at the highest grade, sillimanite (kyanite and sillimanite are aluminosilicate polymorphs with the formula Al_2SiO_5). A similar progressive metamorphic series has been recognized in New England, and in many other places around the world.

The line on a map marking the initial appearance of each of these index minerals is termed an isograd, inasmuch as it denotes nearly constant metamorphic grade. A typical example is illustrated in figure 4.14, an isogradic map of the Upper Peninsula of Michigan and adjacent areas. Although most regional metamorphic terranes display an oriented mineral fabric, these rocks for the most part do not (and thus probably were recrystallized under nearly static load conditions). The position of an isograd, and therefore the metamorphic grade, is related to both temperature and pressure; in the common case where the minerals involved are complex solid solutions, it is also a function of the bulk compositions of the rocks and their included pore fluids.

Not all regional metamorphic belts display the same exact zonal arrangement of index minerals described above. In the metashales and associated rocks of northeastern Japan, for instance, the chlorite zone is only feebly developed, garnet precedes biotite in lower grade rocks, and higher grades are characterized by andalusite (a third Al_2SiO_5 polymorph), cordierite (a complex $Mg-Fe^{+2}-Al$ silicate), and, finally, at highest grades, sillimanite. Kyanite and staurolite are absent, indicating recrystallization at lower pressures than those attending metamorphism in Scotland and New England.

Sodic plagioclase is stable in low-grade metamorphic rocks of both

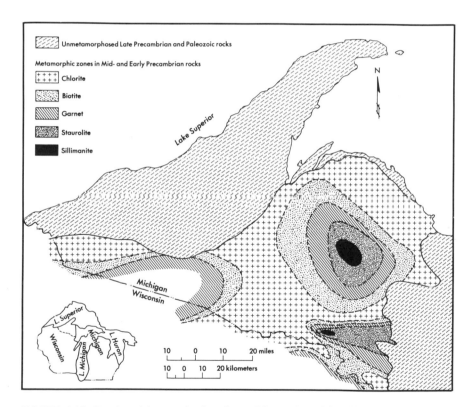

FIGURE 4.14. Regional isograds developed in rocks of the Upper Peninsula of Michigan during Precambrian time (simplified after James 1955). This complex actually shows characteristics transitional between regional and contact metamorphic types. The metamorphic terrane was uplifted, partially eroded, and is now overlain by Late Precambrian and younger supracrustal layered rocks.

types of metamorphic terrane just described. In rocks representing a third regional metamorphic variety, however, this phase may be replaced in the chlorite zone by a pyroxene rich in $NaAlSi_2O_6$ (the mineral jadeite) plus quartz. Higher-grade rocks carry manganese-bearing garnet, which appears before biotite. Another distinctive mineral is the blue amphibole, glaucophane $(Na_2Mg_3Al_2Si_8O_{22}(OH)_2)$. Examples of this association, stable only at high pressures and low temperatures, have been described, for instance, from western California, southwestern Japan, New Caledonia, and the Alps.

The Japanese petrologist Akiho Miyashiro broadly categorized these three distinct types of regional metamorphism as reflecting the following physical conditions: (1) intermediate pressures and temperatures; (2) relatively low pressures and high temperatures; and (3) relatively high pressures and low temperatures. It must be pointed out, however, that every metamorphic belt has recrystallized under its own unique physical regime. In fact, the variations of conditions (in other words, the metamorphic geothermal gradients) also differ from place to place within a specific terrane.

The discussion of regional metamorphism has been concerned so far chiefly with metaclastics. Mafic igneous lithologies also exhibit a distinctive sequence of mineral changes resulting from progressive metamorphism, first documented concisely by the Finnish petrologist Pentti Eskola. Instead of considering individual index minerals, Eskola employed phase assemblages, or metamorphic mineral facies (associations), in rocks of basaltic composition to delineate a progressive metamorphic zonation. The recognition of critical mineral associations has been elaborated on by numerous recent workers.

The common progressive metamorphic sequence of phase assemblages developed in mafic rocks is as follows. The lowest-grades are characterized by complex, low-density, hydrous calcium- or sodium-aluminosilicates belonging to the zeolite mineral group (an example of the open framework structure zeolites possess is illustrated in figure 3.21). These phases occur also as diagenetic minerals, demonstrating the link between sedimentary and metamorphic processes; host rocks are assigned to the zeolite facies. Although relatively volatile-rich, greenschists are more thoroughly reconstituted, compact, typically schistose rocks that contain albite, chlorite, the calcium-amphibole actinolite, and epidote (a hydrous Ca-Al-Fe^{+3} silicate). Where foliation is absent and rocks are massive, these metamorphic lithologies are called greenstones. Intermediate-grade assemblages consisting of the aluminous sodium-bearing calcium-amphibole hornblende, moderately calcic plagioclase, and garnet characterize the amphibolites; a lower-grade portion, in which epidote is stable with albite, can be recognized in some metamorphic belts. The highest-grade mafic rocks are pyroxene granulites, which contain pyroxenes and calcic plagioclase, usually accompanied by garnet. As with the mineral zones developed in metashales, the sequence of metabasalt and metagabbro assemblages reflects increasing pressures and temperatures, proceeding from the most volatile-rich, fine-grained zeolitic rocks and greenschists to the CO_2- and H_2O-poor, coarse-grained pyroxene granulites.

In different regional metamorphic terranes, metabasaltic phase assemblages show certain striking dissimilarities. In areas subjected to relatively low-pressure, high-temperature metamorphism, pyroxene hornfels are widely distributed at the expense of more volatile-rich rocks. The mineral cordierite is common, whereas garnet is nearly lacking. Epidote-plus-hornblende-plus-albite rocks are not well developed. In contrast, where relatively high pressures and low temperatures attended the metamorphic recrystallization, blueschists rather than greenschists occur as low-grade rocks. Here, the mineral assemblages include the blue amphibole, glaucophane; albitic plagioclase is either less abundant or is totally absent. Highest-grade, coarse-grained rocks include eclogites, which contain as principal minerals an intermediate sodium-calcium pyroxene and garnet.

The contrasting sequences of index minerals and phase assemblages developed in metashales and in metabasalts of regional metamorphic belts are presented schematically in figures 4.15, 4.16, and 4.17. Mineral relationships for P-T conditions appropriate to the metamorphism of portions of Scottish Highlands, New England, and Finland are illustrated in figure 4.15, the relatively low-P, high-T sequence characteristic of northeastern Japan in figure 4.16, and the relatively high P, low-T progressive metamorphism of western California, the Alps, and Indonesia in figure 4.17. In these diagrams, relationships between the mineral assemblages of metabasalts and the index minerals developed in metashales are also illustrated. The reader should study the contrasts between minerals developed in rocks of differing chemistry subjected to the same set of physical conditions, and the disparities among the mineral assemblages of a specific lithology from belts characterized by different thermal gradients. The former demonstrates the influence of rock bulk composition on the equilibrium mineral association at a given P and T, whereas the latter reflects the effect of contrasting physical conditions.

What sorts of chemical changes take place during metamorphism? The products of dislocation metamorphism correspond almost exactly in composition to those of protoliths, because granulation does not involve significant chemical reaction and differential movement of elements. On the other hand, where thorough recrystallization attends metamorphism, solutes carried in the attendant fluid may be partially exchanged for constituents in the country rock, and the resultant bulk composition of the metamorphic rock will be altered. This process of chemical replacement, known as metasomatism, occurs to a greater or lesser degree in virtually all recrystallized rocks. It is very important in deep-seated ultrametamorphic and batholithic belts, and is responsible for many ore deposits formed at shallower crustal levels around igneous intrusions (see chapter 6).

The hydration and carbonation of an initially nearly anhydrous mafic volcanic rock is a type of metasomatic alteration. The progressive dehy-

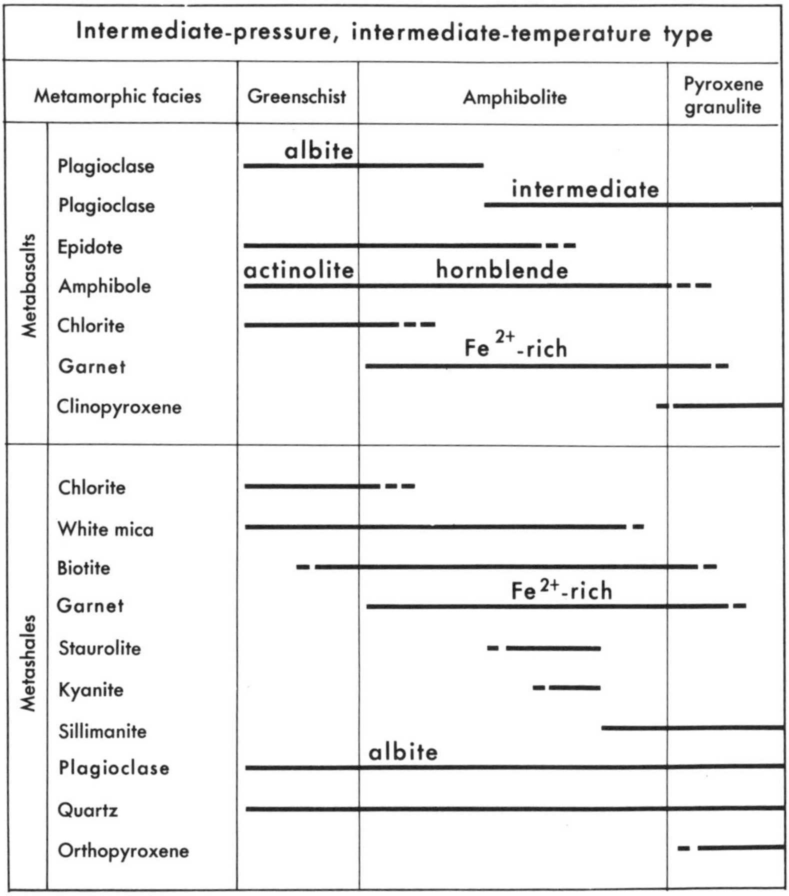

Intermediate-pressure, intermediate-temperature type			
Metamorphic facies	Greenschist	Amphibolite	Pyroxene granulite
Metabasalts Plagioclase	albite	intermediate	
Plagioclase			
Epidote			
Amphibole	actinolite	hornblende	
Chlorite			
Garnet		Fe^{2+}-rich	
Clinopyroxene			
Metashales Chlorite			
White mica			
Biotite		Fe^{2+}-rich	
Garnet			
Staurolite			
Kyanite			
Sillimanite			
Plagioclase	albite		
Quartz			
Orthopyroxene			

FIGURE 4.15. Mineral assemblages in "normal" regional metamorphic terranes (intermediate-P and -T type). Not all the phases shown need be present in any particular rock (Ernst 1969).

dration of a low-grade metamorphic rock such as a metashale or a metabasalt is another example of metasomatism, one in which an aqueous fluid phase is evolved and expelled from the recrystallized mineral assemblage. During the metamorphism of sedimentary rocks under conditions of rising temperature, organic debris is converted into carbonaceous material and, ultimately, into graphite through dehydrogenation. If ferric iron-bearing phases are present, they may be reduced, the oxygen produced being combined with carbon and gradually driven off as CO_2 under successively more intense metamorphic conditions. Oxygen obviously is lost through this mechanism and also by expulsion of SO_2 after combination of O_2 with sulfur (the precursor of which was iron

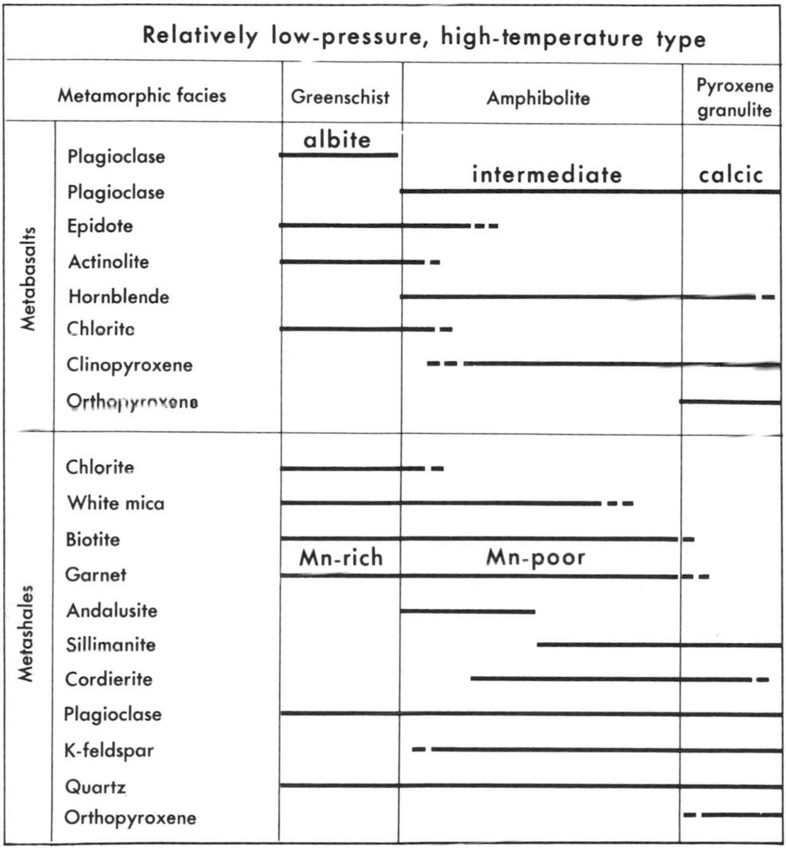

Relatively low-pressure, high-temperature type			
Metamorphic facies	Greenschist	Amphibolite	Pyroxene granulite
Metabasalts Plagioclase	albite	intermediate	calcic
Plagioclase			
Epidote			
Actinolite			
Hornblende			
Chlorite			
Clinopyroxene			
Orthopyroxene			
Metashales Chlorite			
White mica			
Biotite			
Garnet	Mn-rich	Mn-poor	
Andalusite			
Sillimanite			
Cordierite			
Plagioclase			
K-feldspar			
Quartz			
Orthopyroxene			

FIGURE 4.16. Mineral assemblages for metamorphic terranes formed under a relatively high geothermal gradient (relatively low-P and high-T type). Not all the phases shown need be present in any particular rock (Ernst 1969).

sulfide and/or biogenic S_2). The net result is that high-grade metamorphic rocks characteristically are impoverished in volatile constituents relative to low-grade equivalents.

Devolatilization may involve only relatively minor changes in the proportions of nonvolatile constituents of the rock. However, laboratory experiments show that, at high confining pressures, hot aqueous fluids can dissolve large amounts of alkalies and silica, and moderate amounts

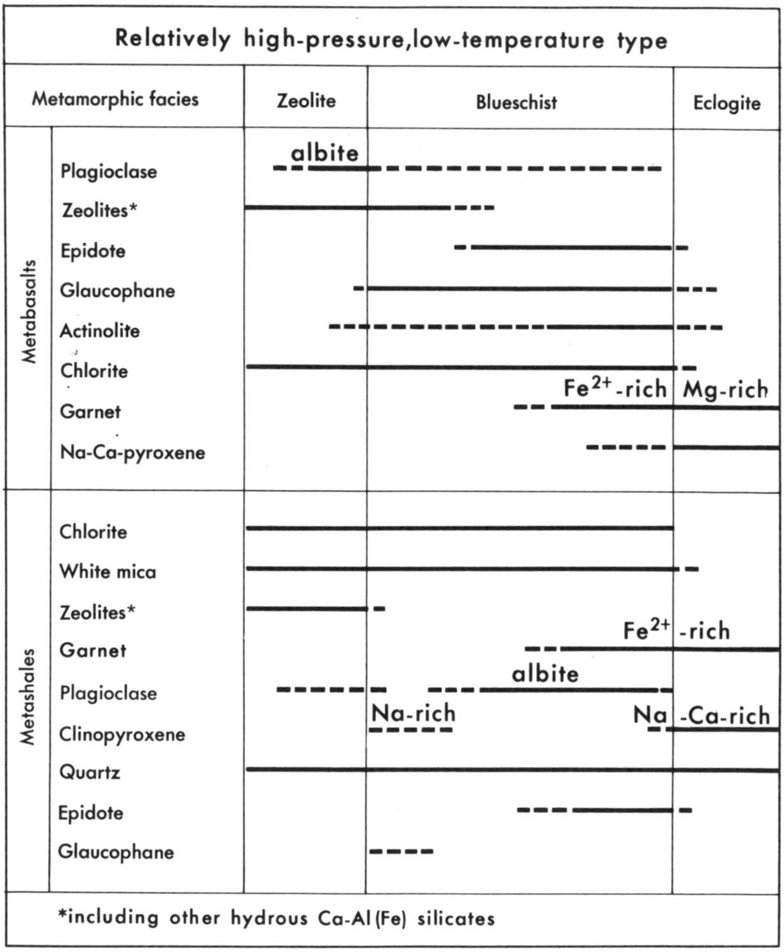

FIGURE 4.17. Mineral assemblages for metamorphic terranes formed under a relatively low geothermal gradient (relatively high-P and low-T type). Not all the phases shown need be present in any particular rock (Ernst 1969).

of other constituents as well. Therefore, under high-grade metamorphic conditions appropriate to the deep continental crust, H_2O-rich fluids generated by dehydration reactions are highly charged with dissolved sodium, potassium, and silicon. As these fluids migrate toward the surface, they react with the overlying, cooler country rocks to produce alkali feldspars and quartz. These rocks are thus converted to more granitic mineralogies and bulk compositions, the metasomatic process being referred to as granitization. Gneissic terranes exposed in the Canadian, Sino-Korean, Brazilian, and Baltic shields (ancient continental nuclei) contain rocks that appear to have been granitized, at least in part, by the process just outlined. We have previously seen that some felsic magmas are generated near the base of thick sections of the continental crust where temperatures have reached the onset of melting in the presence of H_2O. In this same environment, chemically reactive hydrothermal fluids are also generated by metamorphic recrystallization. In fact, metasomatic phenomena commonly are spatially associated with granitic plutons—at least some of the latter being the end result of ultrametamorphism, or partial melting; ultrametamorphism produces migmatite, a mixed, banded rock type characterized by interlayered refractory solid amphibolitic and remelted granitic layers.

At great depths, magmatic and metamorphic processes merge, and because of the profound and complex recrystallization, in some cases accompanied by intense deformation, the origins and histories of the resultant rocks are difficult to decipher. Less ambiguous examples of metasomatism are found in the contact aureoles developed at shallow crustal levels where hydrothermal fissure fillings and alterations of country rocks (such as the replacement of recrystallized limestone by silicate minerals, sulfides, or iron oxides) are apparently caused by intruding magmas. Such magmatic and metamorphic processes are more clearly distinguishable at shallow depths inasmuch as they take place under widely disparate temperatures.

Physical Conditions of Metamorphism

The total pressure attending recrystallization at any specific depth in the Earth is simply the weight of the overlying column of rocks per unit cross section, or load; the pressure is equal to the average specific gravity of the lithologic section times the gravitational constant times the depth. It is easy to compute for any distance below the surface, provided the average specific gravity of the rock column is known. Of course, a knowledge of the correct depth is also required for the calculation, and this usually must be inferred from geologic reconstructions.

Because rocks are weak except at very low temperatures, differential stress evidently does not contribute significantly to the pressures maintained in deeply buried units.

What we need to know is, what was the temperature of metamorphism at a given depth in the Earth? Answers may be obtained in several ways. First, temperature can be estimated by extrapolating present-day geothermal gradients measured in the upper few kilometers of the crust to greater depth. Second, temperature can be approximated through correlation of phase assemblages synthesized in the laboratory under controlled conditions with observed mineral relations in rocks of known, or inferred, depth of crystallization. Finally, because the equilibrium partitioning of stable isotopes (particularly of low-atomic-number elements such as hydrogen, carbon, and oxygen) between coexisting mineral pairs is a function of temperature but is unrelated to pressure, measured isotopic fractionations can indicate the temperature that attended crystallization of the enclosing rock.

Nevertheless, the conditions of metamorphism are imperfectly understood at present. The reasons are that ambiguities exist in interpretation of the geologic record, that laboratory phase-equilibrium experiments are performed principally on simple chemical compositions differing from those of real rocks, and that in many cases in nature neither chemical nor isotopic equilibrium was achieved during the recrystallization. Thus, uncertainties attend the estimates of P and T of metamorphism.

Schematic diagrams for the physical conditions of individual mineral stabilities in metashales and of phase assemblage stabilities in metabasalts (and metagabbros) are presented in figures 4.18 and 4.19, respectively. These interpretive figures attempt to combine the types of geologic and geochemical data described above into an integrated, consistent picture. The actual values of P and T for a specific mineral assemblage or index mineral should be viewed with caution, however, because variation in the bulk composition of the rocks, including metamorphic pore fluid where present, markedly influences the temperature-pressure positions of most of the mineral and metamorphic facies boundaries presented. Because of chemical complexities for virtually all phases involved in the reactions, although the first appearance of an index phase (isogradic concept) may be represented by a line in pressure-temperature coordinates (figure 4.18), the gradual transformation of one mineral assemblage to another (facies concept) must be illustrated as a zone of finite P-T width (figure 4.19).

It is evident that the different mineral sequences and assemblages characterizing specific metamorphic belts reflect disparate metamorphic geothermal gradients. To obtain an appreciation for the reason behind such contrasting P-T trajectories within the Earth, we must return to a consideration of the thermal regimes of particular lithospheric plate environments.

Plate-Tectonic Settings

Assuming the correctness of the crustal and upper mantle thermal structures illustrated previously as figures 4.9 and 4.10, and of the schematic metamorphic petrogenetic grid of figure 4.19 for mafic rocks, the spatial disposition of metamorphic facies in the neighborhood of divergent and convergent plate boundaries can be approximated, as shown in

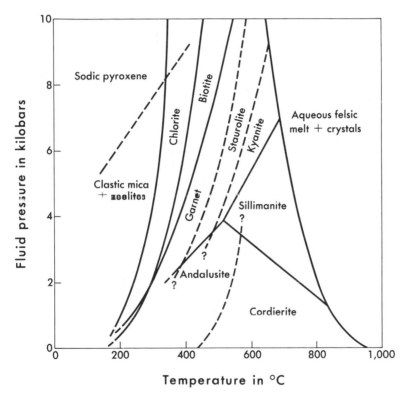

FIGURE 4.18. Pressure-temperature phase diagram for metashale bulk compositions in the presence of an aqueous pore fluid (Ernst 1969). Curves show the first appearance of a phase on the high-temperature side (except for sodic pyroxene, which forms only at relatively elevated pressures). The mineral boundaries, modified from experimental phase-equilibrium studies, correspond to metamorphic isograds shown on geologic maps such as figure 4.14.

figure 4.20 and 4.21, respectively. Isogradic zonations (figure 4.18) would be disposed with somewhat similar geometric relationships. Chemical equilibrium has been assumed. Local departures from the inferred thermal regime, variations in rock bulk compositions, and the incomplete extent to which equilibrium has been attained undoubtedly account for some of the complicated natural occurrences, but the general relationship between recrystallization and lithospheric plate environments seems to be reasonably well understood.

It is apparent from figure 4.20 that a relatively simple metamorphic zonation should be developed in the vicinity of an oceanic ridge. Rocks

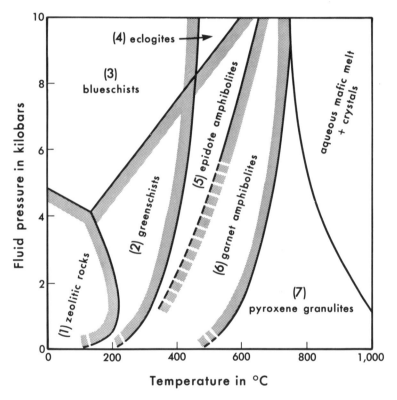

FIGURE 4.19. Pressure-temperature phase diagram for metabasalt bulk compositions in the presence of an aqueous pore fluid (after Ernst 1969). The mineral assemblage boundaries (in reality, pressure-temperature zones much broader than the schematic curves illustrated) are modified from experimental phase-equilibrium studies, and correspond to metamorphic mineral facies boundaries shown on geologic maps. The fields of eclogite and of garnet granulite expand to both higher and lower temperatures where aqueous fluid is absent. Facies numbers correspond to those of figures 4.20 and 4.21.

characteristic of the several high-T, relatively low-P metamorphic facies apparently are formed in the vicinity of the ridge axis, subsequently cool, and are transported laterally away from the source of heat during the sea-floor spreading process. Concomitant with declining temperatures, retrograde (that is, lower metamorphic grade) reactions occur to a greater or lesser degree, depending on chemical kinetics. For any reaction, rates depend on factors such as the accessibility of seawater and extent of granulation, shearing, and fracturing. However, the most important variable is temperature: the higher the T, the more rapid the chemical reaction.

Near a convergent plate junction, similar to the one illustrated schematically in figure 4.21, the nature of the thermal structure evidently produces an oceanward, relatively high-P, low-T metamorphic belt, and a landward, relatively high-T, low-P metamorphic terrane. The former is a narrow, elongate zone of metamorphic rocks that exhibits pronounced asymmetry in mineral assemblages, with highest pressure, most deeply subducted sections lying along the ancient plate junction, and

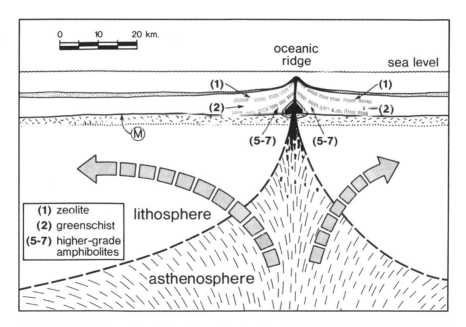

FIGURE 4.20. Schematic distribution of metamorphic facies in basaltic rock types near a divergent plate margin, assuming the general correctness of figures 4.9 and 4.19. Relatively high-T, low-P recrystallization is confined to the vicinity of the mantle plume. The partial fusion of rising asthenosphere to yield oceanic basaltic melt (black) is also shown (after Ernst 1976, figure 6.47).

successively more feebly recrystallized rocks disposed progressively seaward. This prism of rocks is profoundly disrupted by a family of braided thrust faults and shear zones that roughly parallel the convergent plate junction and dip beneath the continental margin or island arc. Tightly squeezed folds are overturned toward the subducting plate and reflect tectonic imbrication (stacking) of the metamorphic section. In contrast, the landward terrane occupies a much broader region and seems to be characterized by a vertically telescoped, relatively high-temperature progressive regional metamorphism, and by a crude bilateral symmetry of mineral zonations about the volcanic-plutonic axis. The arc, of course, represents a positive thermal anomaly, and is typified by elevated heat

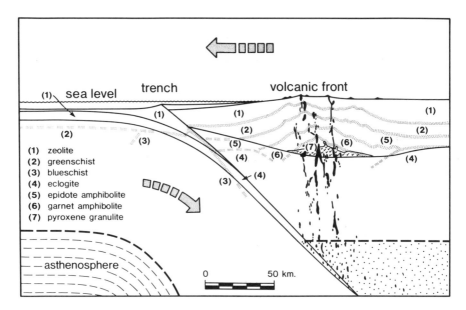

FIGURE 4.21. Schematic distribution of metamorphic facies types near a convergent plate margin, assuming the general correctness of figure 4.10 and 4.19. Relatively high-P, low-T recrystallization takes place in the narrow, asymmetric outboard subduction zone, whereas a broad, bilaterally symmetric zone of high-temperature, low-pressure metamorphism typifies the landward magmatic arc. The partial melting of descending, recrystallized oceanic crust and of relatively undepleted mantle material to provide island-arc or continental-margin basaltic and andesitic/dacitic melts (black), and the partial fusion of H_2O-rich basal portions of the continental crust (checks) are also illustrated. The width of the volcanic belt is only twenty to thirty kilometers at any one time, so in the figure, magmatic activity shown has been integrated over a considerable time span (after Ernst 1976, figure 6.48).

flow. Here faulting is generally high-angle, and folds tend to be more open and gentle compared with the subduction zone.

Thus the three-dimensional geometry of metamorphic belts, and the sequences of mineral assemblages observed within any specific terrane, are complicated functions of the original lithologies and of the P-T regime, as well as the ongoing deformation. The temperatures and pressures attending recrystallization obviously must be related to the large-scale dynamic processes to which the region has been subjected. As with sedimentary and igneous petrology, we may conclude that the areal disposition of metamorphic belts and facies sequences are a function of the plate-tectonic setting.

THE ROCK CYCLE

The erosion cycle discussed earlier in the chapter represents part of an even more extensive reworking of earth materials, the rock cycle. Briefly, the rock cycle relates the history of formation (that is, the genetic lineage) of various lithic units to one another, to their immediate and ultimate sources, and to the continuing compositional differentiation and growth of the crust. As we have seen, this differentiation involves crystal-melt fractionation (including partial fusion) of both uppermost mantle and lowermost crust, differential solution, mechanical and chemical transport, sedimentation, and metasomatism. As will be discussed in the last chapter, theories have been advanced to account for the primary differentiation of what was presumably an initially homogeneous planet into the present layered Earth: metallic iron-plus-nickel alloy core, magnesium- and iron-rich silicate mantle, and alkalic, siliceous crust. Most scenarios propose an early stage of widespread melting of the Earth and sinking of the heavy constituents, rising of the phases of lower specific gravity, followed by a much longer annealing (recrystallization) and cooling period characterized by an ever closer approach to density stability.

This latter stage is the situation in which the Earth appears to be at present. Nevertheless, the generation of silicate melts at considerable—but variable—depths, followed by upward migration, is still going on today. As discussed briefly in connection with the origin of magmas, once the mantle had formed, nearly continuous incipient fusion of rising asthenospheric plumes and sheets beneath spreading centers can account for the worldwide outpourings of basaltic lavas that constitute the oceanic crust. Such simatic materials have been produced throughout geologic time, but, as discussed in chapter 2, most of the oceanic crust has been returned to the deep Earth during the subduction process attending continued convection. The geochemical/lithologic evolution of the Earth will be sketched in chapter 7.

In the mantle beneath the continents, somewhat more andesitic magmas are being generated as a consequence of partial melting in the vicinity of a descending lithospheric plate. However they may originate, primary intermediate and sialic melts that are more silicic and felsic than the parental mafic crust and ultramafic mantle material are being generated currently; furthermore they, too, seem to have been produced episodically through the whole of recorded geologic time. In short, new material is being added to the continental crust semicontinuously.

Whether sialic igneous rocks crystallize at considerable depths or on the surface, many eventually are uplifted, weathered, and eroded. The products of erosion are bodily transported and, after mixing with contrasting materials, are deposited as sediments, then lithified during and after burial. In other environments characterized by the presence of volatiles, the pre-existing igneous, metamorphic, and/or sedimentary

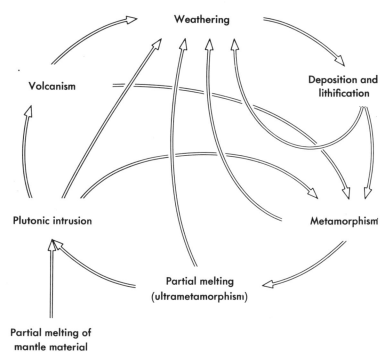

FIGURE 4.22. The rock cycle (Ernst 1969). Different paths and successions characterize the crustal histories of earth materials from different geologic locales.

rocks may be metamorphosed or even partly remelted near the base of the continental crust through the process of ultrametamorphism. The products of these changes are metamorphic rocks and secondary plutonic igneous rocks, respectively.

Both sedimentary and metamorphic rocks are derived ultimately from igneous parents—the primary sources of crustal material. Like igneous rocks, both sedimentary and metamorphic rocks may be subjected to the changes described above; namely (1) erosion and deposition as new sediments, (2) granulation and recrystallization to produce metamorphic rocks, or (3) partial fusion, giving rise to igneous rocks (and residual, refractory metamorphics). This rock cycle is illustrated schematicaly in figure 4.22.

This complex sequence of lithologic addition and reworking is repeated over and over in different orders as required by geologic circumstances. The process results in the increasing heterogeneity and compositional differentiation of the crust, as well as in a gradual increase in the total volume of sial due to the net addition of material from the mantle. The entire cycle is intimately related to the deformational process—itself a consequence of buried heat and consequent mantle flow. Clearly, the rocks and minerals that constitute the dynamic Earth are constantly involved in the petrologic processes of crystallization, transformation, breakdown, and formation anew.

5

THE CIRCUMPACIFIC "RING OF FIRE": EARTHQUAKES AND VOLCANOES

In examining a topographic/bathymetric map of the Pacific Basin, as illustrated in figure 5.1 (see also figure 1.6), one is immediately struck by two major features. First, the margins of the basin are almost universally bounded today by chains of high mountains—the Andes, the Central American, western U.S., and Canadian cordillera, southern Alaska, the Aleutians, the Kamchatka Peninsula, the Kurile Islands, Japan, the Marianas, the Indonesian Archipelago, Tonga-Fiji, and New Zealand. Second, the adjacent off-shore regions are characterized by precipitous submarine slopes leading to the astounding depths (7–10 kilometers) of the oceanic trenches—themselves relatively narrow declivities that parallel the margins of the bordering continents and island arcs.

REGIONAL GEOLOGIC SETTING

Other than the exclusively near-surface earthquakes typifying spreading ridges, the seismicity of the Earth is spatially related in large part to these continental-margin and island-arc features. Shallow-focus earthquakes (0–50 kilometers) occur in the vicinity of the oceanic trenches, and range to more profound hypocentral depths (intermediate focus equals 50–200 kilometers; deep focus equals 200–700 kilometers) as one moves progressively landward across the arc onto the margin and then well within the continental interior. Geographic relations are illustrated in figure 2.12. Based on the revolutionary new tectonic concepts involving sea-floor spreading, global tectonics, and continental drift, we now know that the observed seismicity is a result of differential motion along lithospheric plate boundaries; in general, earthquake foci outline the edges of the plates themselves.

Now consider the on-land geology. Far from the stable continental interiors and Precambrian shields, the Pacific Rim is marked by chiefly Mesozoic and younger orogenic belts coincident with the mountainous topography. Widespread compressional structures, such as thrust faults and overturned (recumbent) folds, provide evidence of pronounced crustal shortening across the edges of the ocean basins. These fold-belt systems are replete with fault-bounded microcontinental slivers and yet more numerous scraps of oceanic crust; such features attest to the drift

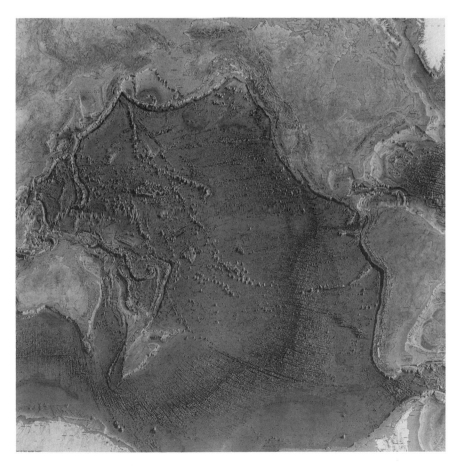

FIGURE 5.1. Bathymetric map of the Pacific Basin, showing major submarine mountain chains and marginal deeps, or trenches (Heezen and Tharp 1977).

of itinerant terranes along the margins of continents and island arcs, and their ultimate stranding against, and suturing to, the accreting sialic margin. A portion of the Mesozoic assembly of North American accreted terranes is illustrated in figure 5.2; also shown is a possible modern accretionary analogue in the Indonesian area of the western Pacific. Many of these orogenic belts are also the site of voluminous andesitic volcanism (see figure 2.16), and where deeply eroded roots of mountain chains are laid bare, huge, subjacent granitic batholiths are exposed.

The above are important manifestations of convergent lithospheric plate motions; they document the progressive constriction of the world's largest basin. The shrinking of the Pacific has taken place as a conse-

(A)   (B)

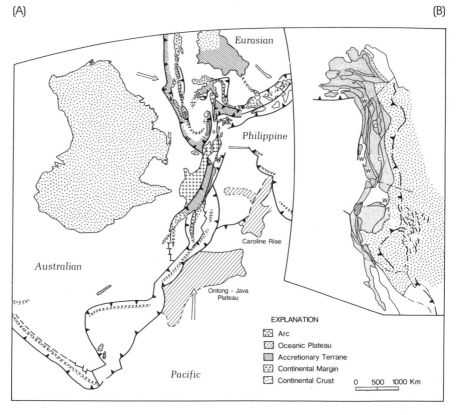

FIGURE 5.2. Mesozoic accreted terranes in western North America (A), compared at approximately the same scale with the convergent suture complex of Indonesia (B), where Pacific, Australian, Eurasian, and Philippine plates are impinging on one another (after Silver and Smith 1983). For a map showing the world's major plates, refer to figure 2.13.

quence of the breakup of Gondwana and the opening (spreading) of the Atlantic Ocean. Sea-floor spreading occurs in the Pacific as well and is reflected by the existence of the Southeast Indian Ocean Ridge–East Pacific Rise system; the velocity of divergence originating along this stupendous midocean ridge complex is six to ten centimeters per year, roughly three times as rapid as the Mid-Atlantic Ridge spreading. However, unlike the Atlantic Basin, the Pacific is characterized by a nearly continuous peripheral ring of subduction zones; and where these are missing, transform plate junctions are present instead (compare figures 2.5 and 5.1). The Circumpacific is the Earth's principal foundry in which new sialic crust is forming and being recycled. Let us therefore examine the reasons why continental growth is occurring along the Pacific Rim. We will concentrate on the geologic relationships of the U.S. West Coast as a typical example and will end with a brief look at the geologic hazards attending evolution of this young, vigorous mountain belt.

⎯ PLATE MOTIONS ALONG THE PACIFIC RIM

The age of ocean-floor sediment lying directly above the basaltic crust of the Pacific Basin is shown in figure 5.3. Chemically precipitated, fossiliferous, and very fine-grained clastic debris settles out slowly but continuously in the deep sea; it begins to be deposited on the oceanic crust immediately after the latter cools and solidifies in the vicinity of a spreading center. Thus, the age of the basal portions of what on figure 2.6 was labeled sedimentary layer 1 (see also figure 2.15) marks the time of final solidification of the underlying basaltic substrate, the so-called layer 2 of oceanographers.

The contrasting ages of sediments resting on oceanic basalts define submarine geographic bands, comparable to those indicated by magnetic anomaly patterns. Both types of "zebra striping" parallel the spreading systems because each is a reflection of the time of formation of the oceanic crust. The oceanic crust and overlying hemipelagic sediments are, of course, youngest at and near the presently active divergent plate junctions, and become progressively older in the directions of spreading orthogonal to the ridge axes as a consequence of the spreading process. Thus, the most ancient sea-floor sediments still preserved in the western Pacific Basin were deposited as recently as Middle Jurassic time, about 180 million years ago. The bilateral symmetry of spreading obviously has been preserved in the Atlantic as mirrored by the ages of the basal sediments (see figure 2.15), whereas in the Pacific, the presumed initial symmetry has been partially obliterated. What has happened? Because of the equivalent amounts of convergent and divergent plate motions required for the Earth as a whole, the geologic record contained in the oceanic crust and its sedimentary cover is fated to be destroyed sooner or later by the subduction process. In the Pacific, the conveyer-beltlike motion of sea-floor spreading brings the crust and very fine-grained overlying strata to a trench at the oceanic basin margin within 100–200 million years as a maximum. Moreover, the rate of plate consumption in the trenches exceeds the production of new lithosphere at the spreading ridges; consequently, the Pacific Basin is gradually shrinking in total area. On the other hand, the Atlantic is still essentially devoid of such subduction zones and so, because of sea-floor spreading, is enlarging.

The Southeast Indian Ocean Ridge lies equidistant between the Gondwana continental masses of Antarctica and Australia–New Zea-

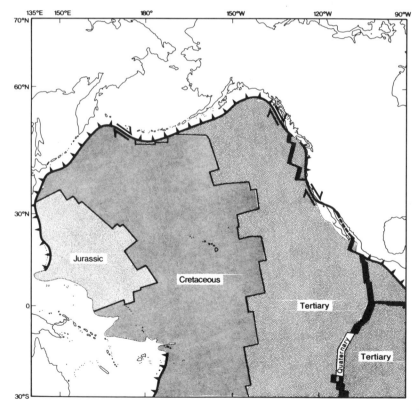

FIGURE 5.3. Age of the basal sediments, and underlying basalts of the oceanic crust in the Pacific Basin, generalized from figure 2.15. Tertiary strata are 66 million years old or less, Jurassic layers are 144 million years old or greater.

land; this spreading system came into existence during Late Cretaceous time (approximately 80 million years ago) when Australia–New Zealand split off from Antarctica, and both drifted northward relative to Antarctica. As the Southeast Indian Ocean Ridge passes into the Pacific Basin, where its extension is known as the East Pacific Rise, it turns northeastward (refer to figures 1.6 and 5.1). The East Pacific Rise impinges on the North American continent in the neighborhood of the Gulf of California, reappearing again north of the Cape Mendocino fracture zone where it is called the Gorda–Juan de Fuca Ridge. These are all segments of the same divergent plate junction, but the spreading ridge has been overrun from the east by the Californian sector of the North American continental crust–capped lithospheric plate.

We can now understand what must have happened to the bilateral symmetry of the Pacific seafloor by reconstructing the Cenozoic movement history. Take the East Pacific Rise as our spatial frame of reference. Today, as in the recent geologic past, eastern lithospheric limbs of the East Pacific Rise move eastward and southeastward, then turn downward and sink beneath the yet more easterly continental crust–capped North and South American lithospheric plates. The modern locus of subduction, of course, is marked by the Chile-Peru and Middle America Trenches; plates moving into these oceanic deeps from the west include the Farallón, Cocos, Nazca, and Antarctic (figure 2.13). However, driven by sea-floor spreading in the Atlantic, the American plates are slowly moving westward, encroaching on these plates that constitute the eastern limbs of the East Pacific Rise. Hence, the convergent lithospheric plate boundaries, bordering North and South America on the west, gradually migrate westward relative to the divergent boundaries in the eastern Pacific Ocean.

Where the two have collided, as in California, trench and ridge are mutually extinguished, and a newly generated transform fault system accommodates the differential motion between the North American and Pacific lithospheric plates. The initial encounter was along the central California coast about 29 million years ago (late Oligocene time). Since then, the collision and transference of plate motions have propagated through western California as pairs of northwest and southeast ridge-trench-transform triple junctions, with the continental transform connecting the two points of impingement where divergent and convergent lithospheric plate boundaries meet. This coming together results in ridge-plus-trench obliteration. The continental transform—a conservative plate boundary—is actually a broad, evolving zone of subparallel strike-slip faults, the most prominent of which is the San Andreas. Three stages in this plate-tectonic scenario, at 37, 20 and 0 million years ago, are depicted in figure 5.4. In this figure, note the "wake" of volcanic islands

(A) 37 Million years ago

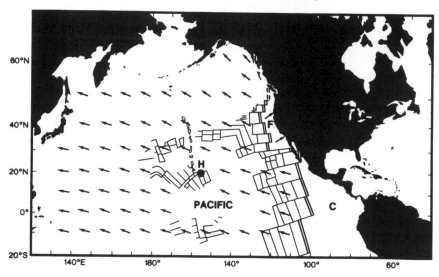

(B) 20 Million years ago

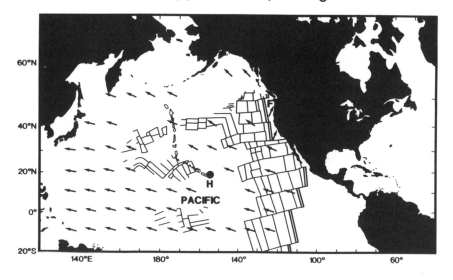

(C) Present

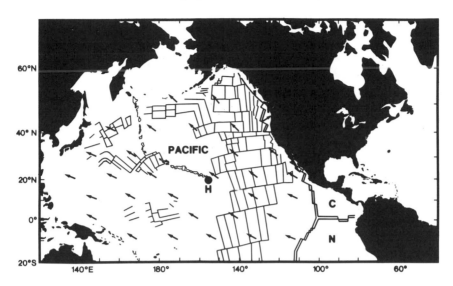

FIGURE 5.4. Glimpses at 37, 20 and 0 million years ago of the dispositions and relative movement directions (vectors) of lithospheric plates in and adjacent to the Pacific Basin (after Engebretson, Cox, and Gordon 1985). The present outline of North America is shown, although prior to 10–20 million years ago Baja was actually snuggled up against the coast of mainland Mexico. Magnetic anomaly patterns are depicted, as is the Hawaiian hot spot (H), and the chain of islands formed by plume eruptions. Prior to about 40 million years ago, the Pacific plate was moving more nearly northward, as indicated by the hot spot track (i.e., the linear island chain). Arrows indicate motion of the plates at the time designated in the figure title. F = Farallón plate; C = Cocos plate; N = Nazca plate.

generated by the Hawaiian hot spot (shown by a large, black dot in the figure). The hot spot is considered as a fixed locus in the deep upper mantle stationed beneath the moving lithosphere. This thermal plume is being overridden by the west-northwestward-drifting Pacific lithospheric plate. The elbow or bend in the island chain was formed nearly 40 million years ago. Immediately prior to this time the Pacific lithosphere was moving in a north-northwestward direction, as indicated by the bearing of these old seamounts in the northernmost Pacific. This relationship provides the answer to the question posed on p. 40.

▭ SEISMICITY: CALIFORNIA AS AN EXAMPLE

We learned in chapter 2 that earthquake hypocenters are concentrated along lithospheric plate boundaries, because these are the portions of the plates subject to differential movement, or shear, and stress concentration (refer to figure 2.12). Within the oceanic realm, a map showing epicentral locations of seismic activity is relatively simple for two reasons: (1) oceanic crust–capped lithosphere is exclusively young, geologically speaking—say, less than 100–200 million years old—and (2), the top of the asthenosphere lies within 50–100 kilometers of the surface. Accordingly, earthquakes mark brittle behavior in rather uniformly thin, homogeneous, ocean crust–capped plates. In contrast, lithospheric slabs surmounted by continental crust may be as much as 200–400 kilometers thick, and being on the average much more ancient—up to 3.8 billion years old!—are riven by numerous discontinuities and zones of weakness reflecting a tortured history of repeated orogeny. Furthermore, the oceanic subduction zones of the world are inclined shear surfaces that dip in general beneath the continents and island arcs. For these reasons, a map of the continents exhibits a much more complicated, smeared-out picture of earthquake epicentral distribution. Nonetheless, plate boundaries and subdomains are also defined, albeit somewhat indistinctly, by on-land earthquakes.

As we have seen, California represents a regional salient where the North American margin has overridden the Pacific spreading system. This geometric configuration is shown diagrammatically in figure 5.5.

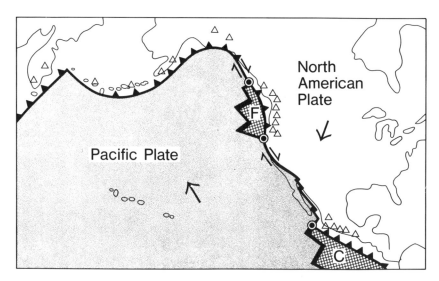

FIGURE 5.5. Present-day plate boundaries of volcanic belts in western North America, after Ingersoll (1982). The correlation of modern andesitic arc volcanism (open triangles) with zones of lithospheric underflow (barbed junctions) is clear. F = Farallón plate; C = Cocos plate.

Thus, although dynamic divergent and convergent plate junctions are situated near the northern and southern extremities of the state, most of the sialic crust in California is transected by a transform system consisting of subparallel northwest-trending strike-slip faults. The San Andreas is the best known, and one of the most active strands, but as figure 5.6 illustrates, it is but one of many major, and innumerable minor, faults within the state.

Large earthquakes are episodic, with intensities roughly proportional to the length of the crustal segment that has been ruptured in a specific seismic event. The Richter scale quantitatively evaluates the amount of vibrational energy released, by referring the quake to the response of a standard seismometer normalized for its distance from the

FIGURE 5.6. Fault map of California. Undoubtedly, many more faults exist than those which have been located on this map. Watch out!

hypocenter. The scale in arbitrary units involves an increase of roughly thirty-two—not ten, as is commonly inferred—between successive numbers: thus, a magnitude 6 earthquake releases approximately thirty times as much seismic energy as a magnitude 5 earthquake, a 7 nearly a thousand times as much as a magnitude 5, and so forth. So-called giant earthquakes have magnitudes of about 7.5 and greater.

Within historic time, devastating quakes have occurred along the San Andreas fault in the vicinity of San Juan Bautista (1800), Cajon Pass (1857), San Francisco (1906), and Loma Prieta (1989). The most severe shaking of all, however, took place in California near Lone Pine (1872) on the eastern Sierra Nevada–Owens Valley fault escarpment. Other major temblors (earthquakes) have struck throughout the state on other faults. Important seismic events include earthquakes at Santa Barbara (1812), Long Beach (1933), Bakersfield (1952), Hayward (1911, 1984), Indio (1968), San Fernando Valley (1971), Coalinga (1983), and Whittier Narrows (1987). Obviously, the San Andreas, although important, is not the only earthquake-prone fault within California. Undoubtedly, many

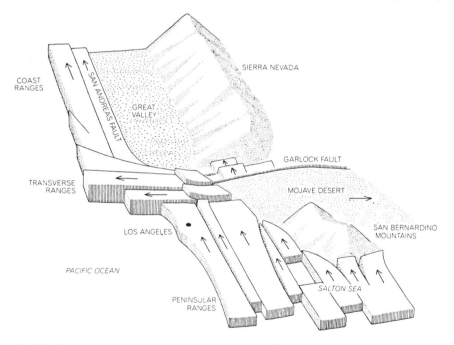

FIGURE 5.7. Very much simplified diagram of differential movement of crustal blocks within southern California along the San Andreas continental transform system (from Anderson 1971). The San Gabriel Mountains are shown as two (unlabled) upthrust blocks west of the Garlock fault and east of the Transverse Ranges.

faults, probably capable of major shaking, have not yet even been recognized, let alone studied.

Most of the seismicity noted above has taken place by horizontal strike-slip motion, as would be expected for transform faulting. This displacement picture is readily accounted for by the lithospheric plate interactions, illustrated schematically in figure 5.7 (see also figures 2.13, 5.3, and 5.5). However, some earthquakes, such as the San Fernando Valley and Whittier Narrows events, involved low-angle thrust faulting. In both of these cases, compression has been (temporarily) relieved through crustal shortening and upward movement of the San Gabriel Mountains block. Why do compression-generated vertical uplifts occur along a predominantly conservative plate boundary? The answer to this question has to do with irregularities in the fault zone.

In general, strike-slip faults are remarkably straight, nearly vertical breaks, as they must be to allow one segment of the Earth's crust to slide horizontally past an adjacent segment. In old continental crust, however, as mentioned in chapter 2, fault bends occur, probably as a reflection of the heterogeneous nature of the sialic crust–capped lithosphere. The heterogeneity can be ascribed to the lithosphere's complex geologic

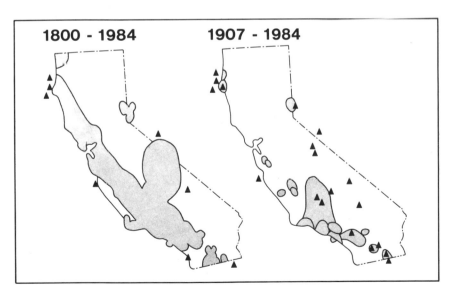

FIGURE 5.8. Regions damaged by seismic activity (stippled pattern) in California since 1800 (after Toppozada, Real, and Parke 1986). Filled triangles indicate earthquakes for which individual, well-defined damage areas are not known accurately.

FIGURE 5.9. Circumpacific earthquake damage due to loss of foundation stability (National Oceanic and Atmospheric Administration photos): (A) leaning and overturned apartment buildings at Niigata, Japan; (B) devastated Government

(C)

(D)

Hill School, Anchorage, Alaska; (C) freeway bridge collapse, San Fernando Valley, California; and (D) total failure of a twenty-one-story office building, Mexico City.

history. Sharp strike-slip fault curvatures are of two types (see figure 2.19): (1) those that open a gap, and therefore are called releasing bends, or pull-aparts; and (2), those that bring excess masses together, and are termed constraining, or compressive, bends. The former occupy the sites of topographic depressions and become basins of sedimentary deposition; the latter result in the upthrusting of segments of the sialic crust, and produce high mountains. The San Gabriel block of the Transverse Ranges represents such a massif (large fault-bounded unit), squeezed upward as the San Andreas experiences a deflection to the west before resuming its more northwesterly course toward and through my residence in Portola Valley, and thence through San Francisco.

As shown in figure 5.8, no area within California is more than a few tens of kilometers from the locus of historic seismic activity, hence the entire state must be considered to be at risk. How earth scientists are attempting to predict earthquakes, and ameliorate their devastating effects will be discussed later in this chapter. For better perspective, we also need to consider California within the broader framework of the Pacific Rim. Ground shaking and loss of life and property have been constant dangers in the past, as evidenced by photos of the devastation at Niigata, Japan (1964), Anchorage, Alaska (1964), San Fernando Valley, California (1971), and Mexico City (1985) presented in figure 5.9. Because of the inexorable differential movements of the lithospheric plates around the Circumpacific, earthquakes will accompany all of us into the future. In order to maximize the chances of survival, therefore, we need to plan for the inevitable "big one."

VOLCANISM IN THE PACIFIC NORTHWEST, CONTERMINOUS UNITED STATES

Modern volcanic activity frames major segments of the Pacific Rim. It is predominantly of the andesitic-dacitic type, and characteristically builds great stratovolcanoes, as in the Andes, High Cascades, Alaska-Aleutian Peninsula, Kamchatka, northeastern Japan (see figure 4.3), North Island, New Zealand, and elsewhere along Circumpacific continental margins and island arcs. Rarely are the extrusives more silicic than dacite, but in many regions (as in the Marianas, the Transmexican volcanic belt, and the Tonga-Kermadec arc), substantial proportions are basaltic andesite and basalt. In all cases, the chiefly andesitic volcanism is sited on stable, nonsubducted plates directly above descending lithospheric slabs. Offshore trenches are the surface manifestations of the convergent plate junctions, with earthquake hypocenters outlining the landward-dipping seismic shear zones. Gaps in andesitic chains are characterized by seaward transform plate junctions (strike-slip motion rather than subduction), such as typify most of California today. You might check for yourself the localization of andesitic volcanoes shown in figure 5.5; in all occurrences, they are sited landward from—and above—the convergent plate junctions, and are conspicuously lacking in the vicinity of transform plate boundaries.

Sialic crust is essentially andesite-dacite in bulk composition; the process that produces this magma type (intrusive batholithic units as well as lavas and pyroclastic deposits) is undoubtedly chiefly responsible for the formation and growth of continents. We hypothesized in chapter 4 that andesitic melts represent the partial fusion products of the subducted oceanic crust that buoyantly ascend from the downgoing slab, or from the nonsubducted mantle wedge overlying the sinking plate, or both. Varying degrees of reaction with the overlying mantle and crust, and commingling with liquids derived from this nonsubducted plate may account for the observed compositional range of igneous rocks that make up the margins of continents and island arcs.

Subduction is occurring beneath the Pacific Northwest, from southernmost British Columbia to northern California. Active volcanoes reflecting this lithospheric underflow constitute the High Cascades, principally situated in western Washington and western Oregon. They are relatively symmetric cones of great beauty, Mounts Ranier and Shasta being two famous examples; their relatively uneroded shapes demonstrate that igneous addition of materials to the superstructures is keeping pace with surficial degradation. Most of the Cascade volcanoes have been regarded as dormant, but the eruptions of Mount Lassen in 1914–15 and Mount St. Helens in 1980 emphatically prove that episodic activity in the High Cascades is an ongoing igneous process.

Every volcanic edifice possessed its own set of responses to the introduction of magma from below. In general, an underlying fracture system transecting the basement acts as the supply route, or conduit, for a rising column of melt. As molten rock enters the superstructure, a volcano tends to inflate slightly because melt is injected as dikes and sills along previously collapsed fissures. Where such tabular igneous bodies reach the surface in a liquid state topographically below the central crater, flank eruptions and satellite cones are constructed. Within the throat of the volcano, the cooling rate may exceed the supply of new magma, and the molten material may congeal, gradually forming a solid plug. Such an impermeable cap traps gases escaping from the cooling melt, resulting in a volatile buildup and eventual rupture and evisceration of the magma chamber as the gases rapidly expand outward and upward, entraining glassy shards (ash) of quenched silicate liquid as well as molten blobs (bombs).

In early 1980, increased heat flow and steam discharge indicated the heightening of igneous activity at Mount St. Helens. Growth of a largely solid dacite plug (and accumulation of underlying magma) generated a prominent bulge at Goat Rocks, on the northern flank of the vent. On the morning of May 18, this blister, a topographically unstable excess mass, was relieved during a landslide triggered by a volcano-induced seismic event of magnitude 5.1 on the Richter scale; the earthquake probably was caused by the upward movement of partly solidified melt in the conduit. Removal of some of the solid overburden material at Goat Rocks allowed a directed, nearly horizontal blast of volatiles and entrained particulate matter to be discharged northward, initially approaching sonic speeds, blowing down forests and wreaking havoc over a 400-square-kilometer area. Relationships are illustrated diagrammatically in figure 5.10, where the sequence of events, and a map view of the altered landscapes are presented.

Figure 5.11 shows some of the damage to a logging camp, providing an indication of the energy released by the directed blast. Depressurization of the magma body was succeeded a few minutes later by a cataclysmic eruption of the contents of the volcanic neck, graphically illustrated in figure 5.12. Much like the rapid evolution of CO_2 from solution when the cap is removed from a warm, shaken bottle of soda pop, the release of expanding gases previously dissolved in the melt carried an ash cloud to stratospheric heights (approximately twenty kilometers). Pyroclastic materials falling near the orifice and immediately downwind accumulated as a thick blanket along the volcanic slopes. Melting snow and rainfall soaked this unconsolidate ash and caused it to rush downhill as mudflows, causing extensive further damage.

The towering column of gas-driven ash, injected into the jet stream of the upper atmosphere, moved east and southeast. Its migration is depicted in the satellite photo of figure 5.13, taken approximately eight hours after onset of the eruption. In all, slightly over one cubic kilometer of magma, and the top of the pre-existing volcanic edifice, were redistributed on the flanks of the cone and to the leeward.

Since the eruptive events of 1980, renewed pyroclastic activity at Mount St. Helens has resulted in the buildup of a small satellite cone within the summit crater of the self-decapitated mountain. Incidentally, an even more spectacular example of this phenomenon is found at Crater Lake, Oregon, where prehistoric Mount Mazama "blew its top" about 6,600 years ago. And, of course, the most famous historic event of this sort destroyed the island of Krakatoa (located between Sumatra and Java) in 1883, killing more than 36,000 people in the associated tsunami, or seismic sea wave. Volcanic particulates—including aerosols—were injected into the stratosphere and circled the Earth for more than two

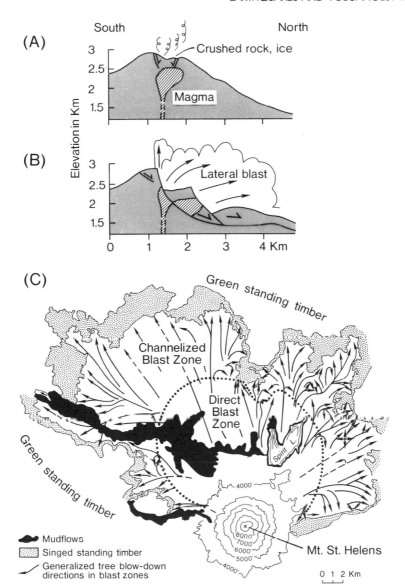

FIGURE 5.10. 1980 eruption of Mount St. Helens, (simplified after Kieffer 1989). The cap preventing upward movement of volatile-bearing magma (A) was relieved by a series of earthquakes and northward-descending landslides (B), in turn giving rise to a directed blast of volcanic gases and entrained debris. This lateral blast moved northward at high speed, initially regardless of obstructions; then, as energy was dissipated, flow became channelized by the local topography (C). Meltwater-soaked ash moved downslope as mudslides, especially westward in the Tootle River Valley.

years, causing brilliant sunsets and a recognizable global lowering of temperatures following the eruption of Krakatoa. Scenarios for nuclear winter find a physical basis of support in such volcanic events.

⸺ GEOLOGIC HAZARD PREDICTION

To reduce the loss of life and to lessen socioeconomic disruption, mankind must be able to predict impending disasters in a timely fashion. This applies to earthquakes, volcanic eruptions, landslides plus avalanches, and related geologic hazards, as well as to other potential natural and man-induced calamities. Only by recognizing the danger can we take preventative action to ameliorate their effects, or to avoid them altogether. Several different, intergradational time scales are im-

FIGURE 5.11. Wrecked logging camp on the north slope of Mount St. Helens. Trucks and stacked timber were strewn about like toys, reflecting the force of the directed volcanic blast (U.S. Geological Survey photo).

portant. Each has associated with it specific opportunities for saving lives and for damage mitigation.

Long-termed hazard prediction—say, on a time scale of years to centuries—provides the possibility of planning land development and site occupancy in a rational manner. Where seismically or volcanically active or avalanche-prone areas are recognized, delineated, and understood scientifically, we can forgo the construction of critical facilities such as hydroelectric or nuclear power plants, petrochemical refineries, or chemical manufacturing facilities. Schools, hospitals, and emergency control centers should not be sited in the vicinity. Of course, all too often such development was initiated long before the hazardous nature of the geologic environment was fully appreciated. Regrettably, at present some new critical facilities are constructed in known hazardous sites anyway, in spite of the documented high-risk site evaluation. Population centers should not expand within geologically dangerous regions, but most do anyway because other considerations, including past history, and climatic, geographic/topographic, socioeconomic, and political advan-

FIGURE 5.12. Eruption of Mount St. Helens. This picture was taken about five hours after the initial landslide at Goat Rocks triggered the northward-directed subhorizontal volcanic blast, in turn directly succeeded by the main-stage pyroxysmal venting of pyroclastic debris shown here (U.S. Geological Survey photo).

tages, are thought to outweigh the perceived hazards. Taipei, Tokyo, Mexico City, Los Angeles, San Francisco, and Santiago are but a few of the world's major, rapidly growing cities at high risk from impending earthquakes, volcanic eruptions, land slippage, or all three. Long-term prediction is of little immediate use in such situations except to upgrade building codes, to reinforce older structures, and to ensure the observation of stringent, safe construction practices in the future. But it does represent a beginning, and, where adequately explained to the population, provides an increase in awareness of natural hazards. A public understanding of the problems is absolutely required for the process of realistic land-use planning to be effective.

Short-term hazard prediction—for instance, on a time scale of hours to days—would allow the evacuation of people and valuable mobile property from dangerous, imperiled geologic areas and man-made structures. Such forecasting requires the timely recognition of unambiguous premonitory phenomena. If prior emergency hazard plans have been proven adequate through simulation testing, orderly withdrawal should be possible, but would not save stationary facilities at risk.

FIGURE 5.13. Satellite weather photo of the fine ash plume from Mount St. Helens, about eight hours after commencement of the eruption (NASA photo).

Long-term prediction is a reality now. The problem is that the likelihood of hazard occurrence in any given, short time period is sufficiently low that populations commonly refuse to provide the financial resources required for research and development in order to mitigate the effects of an inevitable natural calamity in the possibly distant future. Accurate short-term prediction is not yet attainable because of the extremely complicated nature of the geologic hazards themselves. Moreover, predictions themselves may result in severe social and economic disruptions and, if inaccurate, could generate loss of faith by the public, and in a strong political counteraction. Thus, geologic hazard prediction studies are proceeding cautiously and carefully. What sorts of things are being done?

Earthquake frequencies along a particular segment of an active fault may be approximated semiquantitatively employing geomorphic (surface) features in combination with various geologic dating techniques. For instance, escarpments are commonly produced by differential earth movements across an active slip zone; degradation of these scarps by the agents of erosion proceed at observable, measurable rates. Therefore, if several different cliffs are present in contrasting stages of surficial modification, estimates of the earthquake recurrence interval can be made. Similarly, sag ponds are ephemeral features developed along fault traces directly following a major seismic event. The water occupies topographic low spots produced by movement along the faultbreak. Where multiple layers of sedimentary bog deposits occur, the dating of contained carbonized woody fragments in favorable situations may allow an estimation of the local earthquake episodicity to be obtained.

Active faults are being monitored quantitatively employing a variety of techniques. The measurement of differential movements of crustal blocks (including strain buildup within the blocks), fluid and gas contents of fault-bounded rocks, fracture-propagation mechanisms in different lithologic media (both laboratory and field observations), and of numerous other physical parameters are being conducted at the surface, some by Earth-orbiting satellites, and yet others in shallow bore holes. Certain parts of faults slip episodically but rarely, producing great releases of vibrational energy, whereas other portions undergo nearly perpetual creep; both mechanisms are crustal responses to the constant, inexorable, differential motions of the lithospheric plates. Thus, if specific fault segments have been unusually quiet for a long time, or if a fault historically observed to generate an earthquake at periodic time intervals is due, the rough prediction of a future seismic event can be attempted.

Such a situation applies to the Parkfield area in central California, as depicted in figure 5.14. Since 1857, when records began to be maintained in this area, six moderate-intensity earthquakes have occurred, with a roughly twenty-two-year recurrence interval. The last seismic event took place in 1966. As shown in figure 5.15A, the next quake was nominally due in 1988, but has yet to occur. Strain accumulation is continuous at about 2.8 centimeters per year; figure 5.15B illustrates this deformation buildup, as well as the amount of past instantaneous release. Fault slippage during the individual seismic events average about sixty centimeters displacement. Parkfield is heavily instrumented

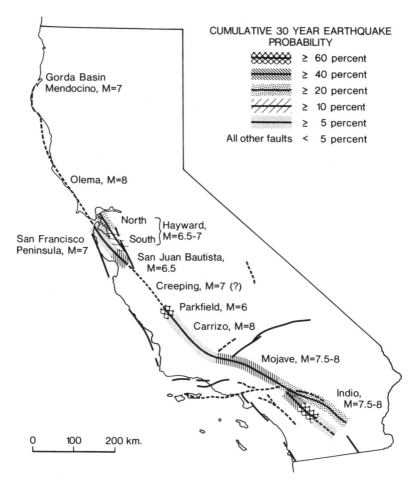

FIGURE 5.14. Cumulative probability of the occurrence of a major earthquake in California over the next thirty years, according to the U.S. Geological Survey. Magnitudes of expected ground shaking for different strands of principal faults are also shown (Richter scale).

and under continuous monitoring because of the expectation that a moderate-intensity quake should occur very soon on this strand of the San Andreas. If precursor phenomena (for instance unusual ground tilts, abrupt changes in water level or gas release in wells, strain buildup in rocks, cracking, and so forth) are detected at Parkfield just prior to energy release, such observations may be utilized there and elsewhere in predicting another impending temblor. However, as of this writing, the anticipated seismic event is already more than a year late.

An increase in the issuance of gases, hot spring activity, and/or measured heat flow near the summit of a volcanic vent, inflation of the volcanic edifice, ground tilting, and volcano-induced seismicity are all indicators suggesting the accumulation of magma beneath and in the conduit, and possible onset of an eruptive cycle. Instrumentation emplaced around Mount St. Helens and Kilauea Volcano on the big island of Hawaii provide case studies of typical andesitic stratovolcanoes and basaltic-shield volcanoes, respectively. The andesitic type is far more explosive, hence dangerous. Recall that viscous andesitic/dacitic melts contain considerable amounts of dissolved volatiles, whereas the hotter, more fluid basaltic magmas are quite low in gassy constituents, hence on eruption, the former blow apart violently whereas the latter extrude quiescently.

Detailed surveying of ground level, downhill movement of soil and bedrock, and abruptly fluctuating water levels in wells may provide warning of the slow-motion creep or impending onset of rapid landslide/

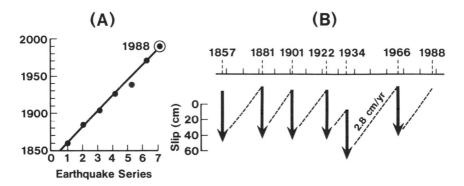

FIGURE 5.15. Historic seismicity of the San Andreas fault near Parkfield, California (U.S. Geological Survey data): (A) earthquake recurrence interval indicated by filled circles (nominal date of next event shown by bull's eye); (B) instantaneous slip during previous earthquakes (heavy downward-pointing arrows) and gradual stress buildup (inclined dashed lines). As of 1990, Parkfield is overdue!

avalanche activity. These premonitory features need to be measured quantitatively and synthesized so that such data may be employed in the prediction of future earth movements which imperil lives and property.

In these and other examples of geologic hazards, earth scientists are endeavoring to quantify the nature of the precursor indicators of an impending major geologic disaster. The hope is to provide accurate intermediate-term and, especially, short-term hazard prediction. Progress toward the realization of these goals is becoming ever more urgent as the Earth's human population burgeons, and as mankind occupies more numerous environments in greater concentrations. Nowhere is this problem more acute than along the Pacific Rim. This is a region of spectacular sociopolitical, economic, and cultural growth. However, as the populations of these nations expand, they are exposed to special environmental risks. The Circumpacific is the region of intensely active mountain building; hence geologic hazards pose a serious, ongoing threat, and one that is heightened as land utilization increases.

═6
═ENERGY AND MINERAL
═RESOURCES

A natural resource is defined as a source of material supporting life, most generally derived from the Earth. The complex, interdependent structure of civilization is built upon a resource base consisting of extractable matter occurring in the biosphere and lithosphere, to a lesser extent, in the hydrosphere and atmosphere. We will confine our attention to Earth materials. Substances utilized provide us with energy and with materials that greatly extend our physical capability to cope with the environment. We may divide these resources into two fundamental categories, abundant and scarce, although, naturally, a spectrum of material concentrations exists. Adjectives such as low-grade and high-grade describe the relative value of the deposit per unit volume or per unit mass.

For most constituents, our primary source of supply is the continental crust. Relative proportions of the major elements are illustrated in figure 6.1. Among the scarce elements, every nation has certain strategic deficiencies—that is essential, required substances in critically short supply. These strategic minerals as they are called, therefore, must be imported; the interruption of such supplies would work a hardship on the importing country and its inhabitants. In any case, importation is a pivotal factor affecting the international trade balance. A picture of the adequacies and inadequacies of U.S. domestic nonfuel mineral resources is presented in figure 6.2. Although most of the substances listed are scarce materials, that is not necessarily the case. As you can see, the U.S. produces only enough aluminum (a major crustal element) from economic deposits to satisfy about 10 percent of our industrial need.

At least for the reasonably abundant elements—say, titanium—the amount residing in the Earth far exceeds mankind's aggregate consumption; what is necessary is sufficient energy to extract the desired substance from the rocks, seawater, or the air itself at a reasonable cost. To minimize the efforts required for mining and refining, anomalously high concentrations of the materials, termed ores or mineral deposits, are sought out for development. Nevertheless, if we are able to afford the energy costs required to extract and beneficiate (i.d., to smelt and refine) them, elements present in remarkably low abundances can be advantageously recovered, employing large-scale mining and concentration methods to enhance cost effectiveness. For instance, the so-called porphyry copper deposits of the western conterminous U.S. are granitoid stocks and batholiths, the rocks of which are successfully mined in large volumes and the ores smelted where the original copper contents are as low as about 0.4 percent by weight!

Scarce elements are present in very low concentrations in typical crustal rocks. Most trace metals, such as lead, occur camouflaged in major phases like feldspar; that is, extremely small amounts of lead replace potassium, sodium, or calcium in appropriate cation sites of the feldspar crystal structure (analogous to the framework presented in figure 3.21). Feldspars can accommodate only trace amounts of lead, but usually the abundance of this element in a rock is so low that the framework silicate is undersaturated with respect to lead. For rocks containing substantially greater lead concentrations, the feldspars and other lead-containing minerals become saturated, hence yet higher lead abundances in the rock will result in the formation of a separate lead-

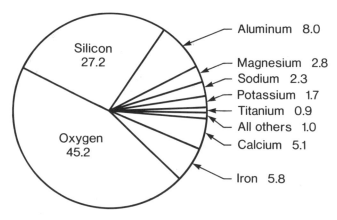

FIGURE 6.1. Percentages of major elements in the sialic crust by weight (Skinner 1986, figure 2.4).

rich phase such as galena (PbS). The cost of recovering lead from galena, understandably, is far less than natural occurrences where lead is widely dispersed as a trace constituent in a mineral of little or no intrinsic value.

The energy required for beneficiation, as a function of concentration, is illustrated schematically in figure 6.3. Accordingly, economic geologists explore for mineral deposits in which rare, sought-after elements are present in excess of the saturation limit—the mineralogical barrier —of common, rock-forming minerals, so that extraction can be performed efficiently and profitably. Not only is the percentage of metal in a typical sulfide or oxide much greater than as a camouflaged trace element in a rock-forming silicate phase, the energy required to isolate the pure metal is generally much less in the former occurrence than the latter.

We can, of course, consider an element—for instance, helium— present in such small concentrations that, no matter how extravagant the energy supplied to the extraction process, once we have used up the

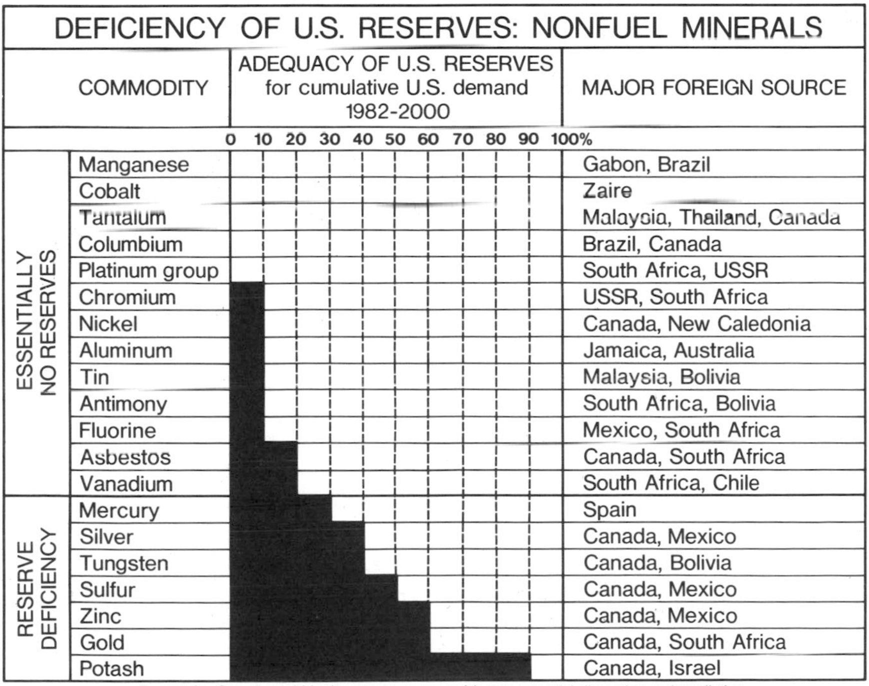

DEFICIENCY OF U.S. RESERVES: NONFUEL MINERALS		
COMMODITY	ADEQUACY OF U.S. RESERVES for cumulative U.S. demand 1982-2000 0 10 20 30 40 50 60 70 80 90 100%	MAJOR FOREIGN SOURCE
ESSENTIALLY NO RESERVES — Manganese		Gabon, Brazil
Cobalt		Zaire
Tantalum		Malaysia, Thailand, Canada
Columbium		Brazil, Canada
Platinum group		South Africa, USSR
Chromium		USSR, South Africa
Nickel		Canada, New Caledonia
Aluminum		Jamaica, Australia
Tin		Malaysia, Bolivia
Antimony		South Africa, Bolivia
Fluorine		Mexico, South Africa
Asbestos		Canada, South Africa
Vanadium		South Africa, Chile
RESERVE DEFICIENCY — Mercury		Spain
Silver		Canada, Mexico
Tungsten		Canada, Bolivia
Sulfur		Canada, Mexico
Zinc		Canada, Mexico
Gold		Canada, South Africa
Potash		Canada, Israel

Except for Canada and Mexico, sources are subject to potential disruption or interdiction.

FIGURE 6.2. Nonfuel dependency of the U.S. on foreign mineral resources (U.S. Geological Survey data).

Earth's entire complement more cannot be obtained because the source is entirely depleted. (In this specific case, future helium needs may have to be met employing the byproduct of nuclear fusion, as described further on.)

Certain forms of energy, such as natural accumulations of hydrocarbons, are produced slowly over the course of geologic time; in contrast, civilization's rate of consumption has skyrocketed within the last few decades. Because oil and gas are present in finite amounts, the accelerating and substantial depletion of this diminishing resource assures that within the first half of the next century, for all practical purposes, terrestrial supplies of oil and natural gas will have been exhausted. When the amount of energy devoted to exploration and extraction of hydrocarbons—or any other form of energy—exceeds the amount of energy recovered in the process, it is obviously no longer sensible to extract this resource. The situation for coal and fissionable fuels (e.g., uranium) are less bleak, but because the supply is limited in these cases as well, future centuries will see a dwindling employment of such forms of energy. In a sense, energy resources represent the fundamental currency of civilization, for our ability to extract and treat all other essential resources requires a sufficiency of low-cost, readily available energy.

At this stage, it is appropriate to distinguish between a mineral or energy resource and a reserve of the same commodity. As we have seen, the former represents the total inventory of the substance residing in the Earth. In contrast, the latter is an economic concept, and refers to that portion of the resource that can justifiably be removed from the ground and refined employing currently available technology. For the free enterprise system, this means in practice that the reserve can be extracted profitably, whereas in a controlled, managed economy, other factors

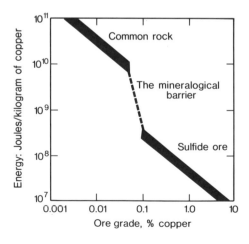

FIGURE 6.3. Energy required (log-log scale) to recover copper metal from common rocks containing minor amounts of copper camouflaged in silicate phases, as well as in copper-rich sulfide ores (after Skinner 1986). The mineralogic barrier is the saturation point of common rock-forming minerals.

such as perceived need—regardless of actual costs—may influence the decision whether or not to exploit a particular deposit. Estimated reserves are best guesses made by knowledgeable professionals, whereas proven reserves reflect amounts of the material which have, in general, been determined by drilling or by some other reliable method documenting the extent and grade of the deposit.

Thus far we have referred to resources of the Earth as if they were limited, and, of course, most are. Certainly only a finite, if large, amount of silicon is sequestered in our planet. So, too, is the situation for fossil and fissionable fuels, as we have just discussed. But although all mineral and most energy supplies are nonrenewable, some minor and one major energy source are renewable, or nearly so. The energy derived from wind, moving water, heat flowing out of the Earth's interior, transformation of biomass, and the solar flux are all examples of energy that is constantly being replenished, at different rates, of course. Some, like wind power and the tides, constitute relatively small potential sources of usable energy, whereas others, such as sunlight, offer major possibilities for future development.

Yet another source of seemingly near-limitless energy is fusion power—the man-made duplication of the process operating within the Sun, whereby hydrogen atoms are combined, or "burned," in an ultra-high-temperature ionized gas, or plasma, to produce helium. Similar to the conversions of biomass, moving air, or flowing water into useful forms of energy, the subject of fusion power is more appropriately one of engineering than it is of geology, so will not be discussed further here. We will concentrate our attention on natural, nonrenewable energy resources of the solid Earth. But as these convenient, cheap energy sources are progressively depleted, civilization will increasingly be obliged to employ, and come to equilibrium with, more nearly constant, renewable sources of energy.

Because society depends so critically on its natural resource base, let us clearly distinguish among several contrasting kinds of usage and growth. Relationships are illustrated diagrammatically in figure 6.4. Exponential growth (A) reflects an ever-accelerating rate of production and consumption, as has been characteristic of our utilization of petroleum and some metals since the turn of the century to about 1960. Incidentally, world human population has exhibited this same form of growth for several centuries. As a renewable resource such as fresh water comes into increasing usage, the rising growth curve must ultimately level off (i.e., establish equilibrium) at the rate at which the resource is replenished (B). Inasmuch as, for nonrenewable resources like oil and natural gas, the total amount of the extractable material is finite, the rising or even exponential growth of production and consumption must inevitably give way to a decline (C), because the total amount of the resource is limited. In all three cases, the amount produced/consumed is proportional to the area beneath the curve. For civilization, the long-term implication of these contrasting growth curves is that ultimately, utilization of energy and other Earth material resources by mankind must achieve a steady state. The implication for population growth is thus clear: at some stage, no growth in consumption is the best we can hope for.

═ HYDROCARBONS, OUR FOSSIL FUELS

Coal

Of the thoroughly exploited energy resources formed by the accumulation and decay of organic matter, coal was the first to be discovered and widely employed. During the Middle Ages, biomass sources—predominantly trees—were burned to provide energy for society, resulting in the denudation of the forests of Europe in the process. Peat and

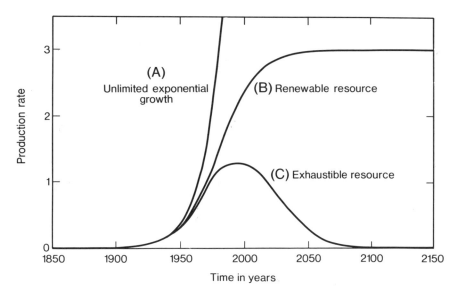

FIGURE 6.4. Schematic growth curves for extraction and/or consumption of a commodity, such as a natural resource (after Hubbert 1985, figure 4.34). Examples from the power industry might include solar and/or fusion energy (A), biomass conversion and/or geothermal energy (B), and petroleum, natural gas, and coal-derived energy (C). The ordinate (left-hand axis) can also be considered to represent a population, inasmuch as numbers of people and total consumption are correlated.

lignite, geologically recent deposits of vegetal matter, then began to be utilized. It was but a small intellectual and technological step to the consumption of coal, and the use of this material of much higher energy content fueled the Industrial Revolution of the nineteenth century.

Coal represents the layer-upon-layer accumulation, decay, and compaction of luxuriant plant matter that initially grew in swamps. Due to later clastic sedimentation, and increasing mass of overburden, these layers were buried to depths sufficient to elevate the rock pressure and temperature; such conditions would tend to drive off most volatile constituents such as hydrogen, oxygen, nitrogen, and sulfur, thereby increasing the carbon content of the resultant bedded organic residuum. Greater depths of burial and the attendant higher P and T, reflecting the weight of the overlying rock column and the Earth's geothermal gradient, thus promote progressively higher-rank coals—that is, materials typified by higher carbon/volatile ratios. The general sequence with increasing pressure and temperature for the rank of vegetal organic matter is peat to lignite to bituminous coal to anthracite. Proceeding to the right, larger recoverable energy values typify the more compressed, devolatilized, higher-rank coals.

Although coal deposits characterize the lagoonal and low-lying continental deposits of many geologic ages, the most widespread and thickest accumulations were produced during the specific Paleozoic period aptly known as the Carboniferous (the so-called Mississippian, and especially, the Pennsylvanian time intervals), about 286–360 million years ago (see figure 7.2). This was a time of nearly ubiquitous, subdued continental relief and widespread maritime, tropical climates; accordingly, exuberant vegetal growth was promoted. Prior to the end of the somewhat earlier Silurian period, plants had not yet occupied ecological environments on land, so terrestrial strata of early Paleozoic ages are devoid of coal deposits (see chapter 7 for a brief discussion of geologic time).

The U.S. is fortunate in possessing abundant deposits of coal of all sorts. Coal fields of the conterminous United States are shown in figure 6.5. Present reserves are sufficient to meet American projected energy needs for several hundred years. Combustion involves the rapid combination of solid carbon with atmospheric oxygen to produce carbon dioxide, liberating large amounts of usable heat in the process. Unfortunately, the carbon dioxide and oxidized sulfur as well as other minor constituents released to the atmosphere by coal-fired power plants, steel mills, and other heavy industries employing the present technologies are resulting in increasing levels of air pollution. Already, acid rain and the CO_2 buildup in the atmosphere and oceans represent serious problems generated by an industrial civilization. Long-term climatic effects, including global warming, are a distinct possibility we must face now. Think what global warming would mean for the Antarctic and Greenland ice caps, for the vast deserts of North Africa, Australia, and central Asia, and for coastal cities around the world.

The problem has been called the greenhouse effect, because it conjures up the heat-trapping characteristics of such glass-walled and glass-roofed structures. Similar to a greenhouse, planetary atmospheres rich in CO_2, N_2O, CH_4, and other gaseous species (e.g., synthetic chlorofluorocarbons) absorb solar radiation; re-emission at different wavelengths prevents efficient transmission back to outer space. The energy absorbance is a reflection of the natural interatomic vibrational frequencies of these atmospheric molecules, which are activated by sunlight. As we burn more fossil fuels, civilization is releasing ever-increasing amounts of carbon dioxide, methane, nitrous oxides, etc. to the atmosphere, changing its composition through the buildup of these molecules and enhancing the greenhouse effect. The only apparent solution seems to involve a long-term decreasing employment of hydrocarbons and other forms of biomass as energy sources.

- EXPLANATION -

Anthracite and semianthracite
Low-volatile bituminous coal
Medium- and high-volatile bituminous coal
Subbituminous coal
Lignite

0   200   400   600
miles

FIGURE 6.5. Coal fields of the conterminous United States (from Averitt 1975). Eastern coals are largely from Mississippian and Pennsylvanian aged sedimentary rocks, whereas western coals, generally of lower rank, occur predominantly in Cretaceous and Cenozoic strata.

Petroleum

More recently brought into usage than coal, petroleum consumption has outstripped this earlier energy supply as mankind's number-one fuel in the last half of the twentieth century. Rates of power generation, worldwide, from coal and lignite versus crude oil since 1800 are shown in figure 6.6. The conversion to petroleum as a fuel has been even more dramatic for the United States. In 1978, the percentages for American energy consumption, as determined by the National Academy of Sciences, were as follows: coal, 18; natural gas, 25; liquid petroleum, 48; hydroelectric and nuclear power, 4 each; all other energy sources, 1. Clearly, for a finite Earth resource such as petroleum, consumption of this magnitude cannot continue indefinitely. As a matter of fact, as figure 6.7 suggests, American petroleum production has already begun its inevitable decline in the lower forty-eight states.

Oil and gas form from the degradation of chiefly single-cell aquatic organisms. In general, rich accumulations of organic matter occur in relatively quiet, anoxic (O_2-poor) waters characterized by slow deposi-

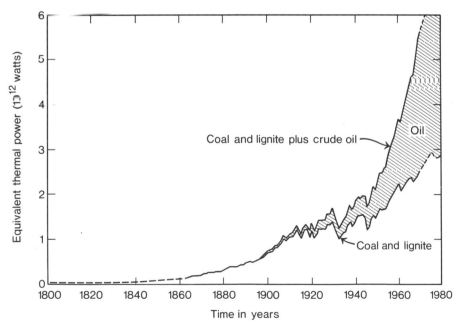

FIGURE 6.6. Worldwide production of fossil fuels (modified from Hubbert 1974, figure 7). The figure illustrates the rapid increase in generation of power from liquid hydrocarbons.

tion of fine-grained clastic and chemically precipitated materials, well removed from oxygenated near-surface waters. Thus the organic material is not attacked chemically through oxidation, nor scavenged by living organisms. Gradually buried by later sediments, the compression and heating that attend continued basinal deposition allow the thermal maturation or distillation and breakdown of the organic material into liquid and gaseous petroleum products. Gas persists to somewhat higher temperatures than liquid hydrocarbons. Although volumetrically rare, petroleum is a characteristic interstitial constituent of lake beds and of many marine sedimentary basins that have undergone substantial burial, but which have not been subjected to such high temperatures that the hydrocarbons have been driven off along with other volatiles during metamorphic recrystallization. At very great depths, of course, the chemical destruction or upward escape of oil and gas is complete; drilling for petroleum is therefore confined to shallower portions of the continental shelves and crust (i.e., the upper five to ten kilometers).

As with ore deposits, only those liquid and gaseous hydrocarbons that occur in significant local concentrations, or oil pools, justify extraction efforts. Because the specific gravity of water slightly exceeds that of oil and gas, with which H_2O is immiscible (incompatible for mixing), hydrocarbons rise to the top of an aqueous fluid column. This difference in density provides the mechanism for the natural migration of petroleum and the creation of "reservoirs," concentrations of petroleum in

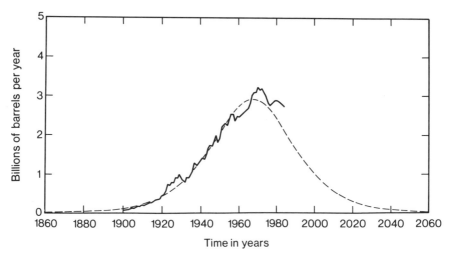

FIGURE 6.7. Crude-oil production in the conterminous United States, superimposed on a mathematically derived smooth curve based on production data (modified from Hubbert 1978, figure 14).

porous rocks. Here, oil and gas occur at levels above those rocks saturated with water. Two different kinds of reservoirs are recognized: structural traps and stratigraphic traps. They are called traps because the
porous reservoir rock (normally a sandstone) containing the petroleum
is overlain by an impermeable cap rock (normally a shale) that prevents
escape of the oil and gas.

Figure 6.8 shows schematic examples of both types. Structural traps
generally involve structural closure on a fold: typically an anticline (A)
rather than a syncline (B), unless water is absent from the pore spaces in
the reservoir rock. Much less commonly it is a fault (C) providing closure. The stratigraphic variety of trap involves a lateral decrease to a
feather edge of the reservoir stratum (D) due to original depositional
features, or variable degrees of cementation. In all cases, the oil and gas
are isolated above the water. As you might surmise, the upward stratifi-

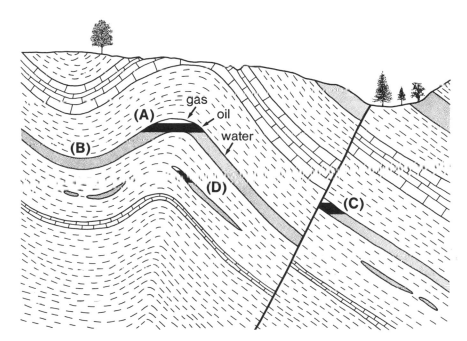

FIGURE 6.8. Schematic cross section of porous, hydrocarbon-bearing sandstones (stippled pattern) interlayered with limestones (brick pattern) and impermeable shales (dashed pattern). The strata are folded and faulted; in addition, some lenses of sandstone occur surrounded by impervious shale. Gas,
being least dense, occurs in voids between the sand grains at high levels in
structural and stratigraphic traps, oil next, and water—the dominant interstratal
fluid—being the densest, occupies interstices in the sandstone reservoir at lower
portions of the section.

cation of water, oil, and gas in the pores of a reservoir rock is a reflection of their decreasing specific gravities.

To summarize, petroleum is generated in fine-grained shaly strata by the decomposition of minute organisms, is partly or completely expelled during compaction and cementation, and accumulates in porous sandy strata where it may be preserved for geologically long periods of time.

Tar Sands and Oil Shales

Vast, near-surface fossil fuel resources occur as extensive but more dispersed and less tractable, low-grade hydrocarbon deposits, such as tar sands and oil shales. The origin of the petroleum in these sedimentary strata is similar to that described in the last section. Pilot plant operations have demonstrated the feasibility of energy recovery, but neither tar sands nor oil shales are economic prospects at present because of technical-development and beneficiation problems as well as formidable environmental concerns. Basically, large amounts of energy must be expended, using present methods, to recover hydrocarbon energy from these low-grade deposits. Because the energy input/output ratio is high, current-day profitability is low to nonexistent. Nonetheless, these enormous energy reserves represent "money in the bank," to be exploited in the future as technology improves. Just what are the natures of these deposits?

The giant Athabaska deposits of northern Alberta constitute the most famous example of tar sands in the world. Here, immense volumes of petroliferous sandstone reservoir rocks have been uncapped by erosion, and subjected to near-surface oxidation and devolatilization. The resultant tarry residue occurs as a viscous, plastic semisolid coating the clastic grains of the sandstone. Because of its stiff, viscous mechanical behavior, the tar cannot be separated easily from the reservoir strata, hence the rocks themselves must be mined before the tar can be extracted. The latter, having lost nearly all its volatiles, is relatively low in energy content, and refining to produce a small fraction of high-energy hydrocarbon fuel is an expensive business.

Oil shales, such as occur in the Green River Basin of northern Utah, northwestern Colorado, and southwestern Wyoming, are widespread, very fine-grained sedimentary deposits characterized by a high content of immature petroleumlike substances. The rocks have not been as deeply buried or heated as most petroliferous strata; accordingly, thermal maturation of the contained organic materials has been less extensive. Moreover, the hydrocarbons are still present in the source shales. In this type of occurrence, similar to tar sands, it is difficult to separate the hydrocarbon from the host rock, but in the case of oil shale, the organic matter occurs as a liquid of moderately low viscosity. It is trapped in the shale because, although porous, this type of rock is nearly impermeable. Thus, as in the tar sand situation, the oil shale must be either mined and retorted (heated to drive off and thermally upgrade the petroleum) at the surface, or else retorted in place by underground combustion. In either case, the process is energy-intensive, hence both expensive and inefficient, and thus far, not feasible from an economic standpoint.

NUCLEAR ENERGY

Nuclear energy refers to that energy released when heavy elements (high atomic number, or Z) such as uranium, split into two or more lighter atoms and elementary particles, or when light elements (low Z) such as hydrogen, combine to form a heavier daughter product. The former process is called fission, the latter, fusion. Both effects are a consequence of the fact that for the intermediate, highly stable elements, with Z between about 8 and 82, the binding energies (which hold an atom together) per atomic mass unit exceed those of the very light and the very heavy atoms by large amounts. Fusion power requires an abundant source of hydrogen, but no resource problem exists for a planet such as Earth, blessed with a globe-encircling hydrosphere. Fission power depends on the availability of the much scarcer heavy radioactive elements such as uranium and thorium. Uranium-bearing minerals constitute the most important terrestrial resources (and reserves) of these elements.

In the crust, uranium cations are present as the two different valence states $+4$ and $+6$. The chief primary concentrations of uranium are associated with felsic rather than mafic igneous rocks, where uranium occurs as the oxide mineral uraninite, UO_2 (also known as pitchblende); minor thorium also occurs in solid solution in this phase where it substitutes for uranium. The uraninite is disseminated throughout the entire rock, is segregated in genetically related lenses of coarse-grained granitic pegmatite, or is present as lower-temperature, post-magmatic fracture fillings (veins) due to precipitation from hydrothermal fluids.

Disseminated uranium in felsic igneous rocks and veins, or their sedimentary detritus, subsequently may be leached during oxidation by near-surface ground water, inasmuch as U^{+6} is much more soluble than U^{+4}. As long as the aqueous phase remains oxidized, the uranium stays in solution. However, if the groundwater encounters organic matter or sulfides as it percolates through a stack of sediments, local reduction occurs in the fluid, and precipitation of U^{+4}-bearing mineral, such as uraninite or carnotite (a complex hydrated uranium-vanadium oxide) may take place. Thus, the important secondary uranium ore deposits of the Colorado Plateau consist of features such as buried, petrified logs that have been partly replaced by uranium-rich minerals. Organic shales and coal beds can produce the same effect.

Pitchblende also occurs as pebbles, sometimes associated with nuggets of gold in Archean (Early Precambrian) stream gravels, as in South Africa and Ontario. These sedimentary deposits, one of a general class of heavy mineral concentrates in gravels known as placers, are a result of the settling out of especially dense clastic grains in the rapid-running axial portions of stream channels, where they are mixed with rocky gravel of much larger grain size but lower density. Specific gravities are as follows: uraninite $= 11$; gold $= 19$; common rocks $= 2.5$–3.0. Detrital uraninite grains appear to be confined to the Archean, and you may wonder why UO_2 pebbles were not oxidized and dissolved. Although the subject is controversial, the answer seems to be that—unlike the present oxygen-rich, near-surface environment—the early Earth was characterized by a strongly reducing atmosphere, and therefore conditions included oxygen-deficient surface waters (see chapter 7). Thus, in ancient times, streams could carry along uraninite particles without dissolving them as would happen today.

⎓ GEOTHERMAL POWER

Thus far we have dealt with nonrenewable energy resources. Heat from the interior of the Earth is produced today largely by radioactive decay of unstable, heavy (high Z) atoms, but in the early stages of development of the Earth, it was also generated by meteorite impact and by core formation (see chapters 1 and 7). This thermal energy is migrating toward the cooling surface at the interface between the solid Earth and both the hydrosphere and the atmosphere. Such heat transfer, involving an almost imperceptible flow of energy, is slow and of small magnitude, but is continuous. This thermal flow is at least partly responsible for mantle convection, plate-tectonic processes, and the formation and evolution of the continental crust. Anomalous concentrations of geothermal energy as it is called, are relatively small, but most are constantly replenished on the time scale of civilization and hence this form of energy may be considered in many cases as renewable.

The trick is to locate sufficiently extensive high heat-flow areas so that exploitation provides a net recovery of energy. Recently active igneous regions constitute the principal geothermal power resource. In favorable occurrences, molten or hot, solid rocks lie close enough to the surface that heat transfer (typically the energy is transported by flowing H_2O) is technologically feasible, yet the igneous heat supply is buried deeply enough to be safe. Volcanic eruptions release enormous amounts of energy, but rather abruptly and often capriciously (chapter 5), hence such violent sources are not easily harnessed at the present time.

Important geothermal power plants have been developed in rhyolitic and basaltic volcanic zones and in active andesitic arc terranes of central Italy, North Island, New Zealand, Japan, Iceland, and the California Coast Ranges. In all these areas, hot springs, steam vents, and geysers spectacularly attest to the transfer of magmatic heat to ground water. In order to harness this power, geysers have been capped by man-made plumbing systems and the naturally produced steam used to drive turbines for generating electricity. A similar development is planned for the basalt dike–intruded Salton Trough at the head of the Gulf of California; here the East Pacific Rise, an oceanic spreading center, reaches a triple junction plate boundary with the overriding North American plate (see figure 2.13 and 5.5), thus accounting for the locally high heat flow. Yet another type of thermal anomaly involves hot, dry rock in an extensional plate-tectonic setting. An example is the Rio Grande Rift of northern New Mexico (figure 2.18), where incipient rifting of the continental crust capped lithospheric plate has allowed the emplacement of now-solidified igneous rocks directly below and on the surface. A tremendous volume of hot, dry rock exists at shallow depths, so heat exchange is effected by drilling wells to moderate depth, pumping cold water down, then recovering the heated effluent in order to drive turbines and to heat buildings.

⎯ MINERAL DEPOSITS

As discussed in the introduction to this chapter, the elements and compounds that sustain life and civilization on Earth are of two inter-gradational types, abundant and scarce. Ore deposits represent unusually high concentrations of the desired substance. For the abundant resources, this means that the local proportion of the material must approach that of the ore mineral itself. For example, the iron ores hematite, Fe_2O_3, and magnetite, Fe_3O_4, occur as nearly monomineralic oxide ore deposits in secondarily enriched sedimentary banded iron formations. In these concentrated occurrences, the content of iron (as metal) may approach 60 or even 70 percent by weight. (The origin of banded iron formations, which were laid down predominantly in Early Proterozoic time, will be briefly discussed in chapter 7.) Quartz sands and quartzites, sources of glass sand silica, may be essentially 100 percent SiO_2. The hydrated aluminum oxides, chiefly bauxite, the major ore of aluminum, can in favorable cases approach pure aluminum hydroxide. The former are reworked, multicycle sandstones in which all the original clastic grains except for the especially hard, nonfracturing and chemically resistant mineral quartz have been degraded and removed as fine particulate matter or chemical species in aqueous solution. The latter are the products of deep tropical weathering, and are termed laterites; the constant, high volume of rain, percolating into and circulating through the rock as groundwater, has dissolved almost all the constituents of the original aluminum-bearing lithologies, leaving behind only the least soluble residue—aluminum hydroxides.

Of intermediate abundances are the elements calcium, sodium, potassium, magnesium, titanium, and manganese. These are relatively abundant and are each concentrated in certain sedimentary rocks of the continental crust; manganese, in addition, occurs as enriched coatings and nodules on the deep-sea floor. Found as derived beach sand placers, titanium is also especially abundant in certain mafic plutons. At the present levels of usage, the Earth as a whole has sufficient reserves of these constituents to accommodate the requirements of industrialized society indefinitely.

Among the scarce metals, copper, lead, zinc, and nickel are the most abundant and are intensively utilized. Here, sulfides constitute the primary ore minerals found in economic deposits. Some metals, such as copper, occur as minor sulfide phases in the shallow parts of granitic batholiths, the porphyry copper deposits of the western United States, Canada, and Mexico (figure 6.9). These plutons, largely of late Mesozoic and early Tertiary age, are confined to the cordilleran (Rocky Mountain) compressional-fold belt and the Great Basin. Nickel is associated with immiscible sulfide droplets that form in mafic intrusive rocks, particularly gabbros, as at Sudbury, Ontario. Lead and zinc form some of their most extractable, profitable deposits as low-temperature sulfide vein fillings and replacements in limestones, as in the central Mississippi Valley; however, PbS and ZnS occur in higher-temperature deposits of various sorts as well. In all these cases, extraction methods involve large open pits (copper) or underground mines (nickel, lead, zinc). The ore is first milled to concentrate the metallic sulfide minerals and then smelted (roasted) to separate the metal from the useful byproduct, sulfur.

Some of these, and other, yet scarcer elements such as the platinum group metals, are being mined today at remarkably low concentrations. The future will undoubtedly see the implementation of even more efficient exploration, mining, and beneficiation methods as the readily exploitable, higher-grade deposits become exhausted. Over the long term, our efforts to utilize these scarce resources will probably be limited by

(1) the finite availability of rare constituents at any concentration level, (2) the energy cost of recovery, (3) the mineralogical barrier (see figure 6.3), and (4), concern for the deleterious environmental impact involved in exploitation, especially where enormous volumes of low-grade material must be mined, transported, and processed in order to extract the desired elements.

INDUSTRIAL MATERIALS

We are all aware that mankind relies on resources typified by high intrinsic values such as petroleum, zinc, aluminum, iron, manganese, chromium, and silver, and this is, of course, correct. However, we employ truly enormous quantities of commoner earth materials as well, often in somewhat less modified form, for many important practical purposes. These include gravel, limestone, ornamental and building stone for the construction industry, mineral fibers and diatomite (siliceous shells of unicellular organisms) for a host of thermal insulation applications and as filters, catalysts, liquid and paper extenders, fertilizers for agriculture, sulfur and other minerals—including petrochemicals—for the chemical and plastic industries, clays and feldspars for ceramics,

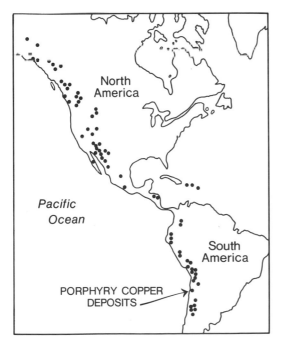

FIGURE 6.9. Late Mesozoic and early Tertiary age porphyry copper deposits of the Cordilleran orogenic chain, western North and South America (after Skinner 1986, figure 6.11). The association with young mountain belts and with continental-crust—capped convergent plate boundaries is apparent.

and so forth. Even more importantly, life-giving water should not be ignored in our treatment of mineral resources. It is required in ever-increasing amounts for industrial, agricultural, and private-sector consumption. The industrial rocks and minerals, and water, are all relatively abundant, readily obtainable, easily extractable materials, but sufficient amounts are absolutely crucial for the development and sustenance of an advanced society. In a sense, urbanization and industrialization depend on a nearby, adequate supply of gravel, among other factors. The point is, civilization utilizes the natural resources available to it, but demands a stable, convenient input of energy, water, and other essential earth materials. Although we have very restricted abilities when it comes to anticipating future technological adaptations and commodity needs, it makes good sense to conserve these, and the scarcer resources, carefully for succeeding generations.

⎓ FUTURE RESOURCE USAGE

Civilization has evolved toward greater and greater usage of an increasingly broad range of earth materials. This trend is most marked in the resource consumption patterns of the industrialized nations—and America is first among the most profligate. As underdeveloped and Third World countries struggle to join the mainstream, they will increase their per capita usage towards current Western levels; thus, as global population continues to rise, and even if world population leveled off, it is inevitable that, for the foreseeable future, earth resources will be under severe stress due to efforts to accommodate the accelerating global demand for minerals extraction and production.

Most near-surface portions of the continental crust have been thoroughly explored and exploited with increasing efficiency. Modern technology will be required to search deeper in our environment in order to obtain the necessary earth materials required to satisfy the voracious appetites of modern societies. This may ultimately mean detailed exploration of the lower continental crust, seabed mining, resource development of rigorous climatic zones such as subglacial, ice-covered Antarctica and Greenland, and extraction of minerals and fresh water from the salty ocean. Mining operations on the Moon, Earth-orbit-crossing asteroids, and even our planetary neighbors Venus and Mars sound fantastic now, but might eventually be feasible as our capabilities in space improve—and necessary if mankind's voracious appetite for raw materials does not abate. Will we learn to live within our terrestrial means?

Modern, technologically sophisticated mining efforts are energy intensive, hence demand a universally available, abundant supply of inexpensive fuel. Fusion and, ultimately, solar power seem capable of providing the long-term solution to our energy needs. But, as world population continues upward, and as man's physical influence over the environment grows, the ability to effect permanent, irrevocable change on our planet is enhanced. The biosphere is a thin, fragile membrane representing intersections with, and couplings among, hydrosphere, atmosphere, and the solid Earth. We cannot predict what the future will bring, but it is absolutely clear that, as more and more resources are consumed at an even greater rate, devotion to the preservation and protection of this almost miraculous, life sustaining zone must become society's highest priority. The future inhabitants of our planet will have to make a living too.

≡7
≡ GEOLOGIC TIME AND EARTH
≡ HISTORY

Condensation of the solar nebula and the accretionary formation of the Earth from planetesimals occurred about 4.5–4.6 billion years ago. This time of origin was determined through the investigation of meteorites and lunar samples, for no terrestrial rocks more ancient than 3.8 (possibly 3.9) billion years old have yet been recognized. By what means were such ages obtained, and how reliable are they? It is in the discipline of geochronology that answers are to be found through the measurement of radioactive isotopes in minerals and rocks. As will now be explained, such systems behave as "atomic clocks."

As a reflection of its particular crystal structure, every mineral contains certain elements in major proportions, others as minor or trace (camouflaged) constituents; still other elements are, for all practical purposes, entirely lacking. Potassium-bearing micas—muscovite, $KAl_2Si_3AlO_{10}(OH)_2$, for instance—contain essential amounts of potassium, but not argon, because of the extreme rarity in condensed matter of this noble gas. Furthermore, a positively charged cation like K^+ is strongly preferred in the appropriate cation structural site over a neutral, nonbonding inert gas atom such as argon.

No matter what the local concentration of potassium, its isotopes occur in similar proportions everywhere on the Earth and, apparently, in the same proportions throughout the solar system. One of these isotopes, ^{40}K (superior figure equals atomic mass, a quantity, specifying a particular isotope of the element) is radioactive. In about 1.5 billion years, therefore, half of the ^{40}K originally present will have decayed to daughter products (including argon) at a measured and well-known rate proportional to the amount of the parent isotope. Then, in the next 1.5 billion years, half of the surviving ^{40}K will have been transmuted, and so on. This gradual, exponential decay and the concept of half-lives are illustrated in figure 7.1. The daughter particles are radiogenic ^{40}Ar (produced by electron capture: proton plus electron yields neutron), and ^{40}Ca (produced by β emission: neutron yields proton plus electron). These daughter particles (along with radiation energy) are generated at a known rate in a fixed branching ratio from the parental ^{40}K. Because radiogenic argon (also ^{40}Ca, of course) is retained in the host mica in the cation site formerly occupied by the parental potassium, the age of crystallization of this phase—hence that of the enclosing rock—can be computed. This is accomplished by measuring the amounts of surviving radioactive parental ^{40}K and accumulated radiogenic daughter ^{40}Ar now present: the higher the $^{40}Ar/^{40}K$ ratio, the older the mica. What has been assumed is that the age of the mica is equivalent to the age of the host rock; neither potassium nor argon have diffused in or out of the structure since the time of crystallization, and the original concentration of ^{40}Ar was zero in the mica.

Another mineral, zircon, $ZrSiO_4$, incorporates trace amounts of ura-

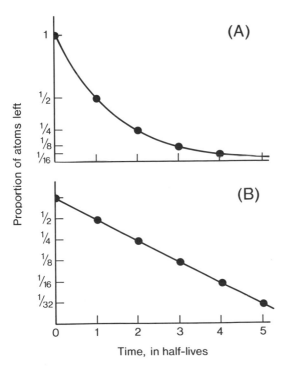

FIGURE 7.1. The exponential decay of a radioactive element, showing several half-lives: each would be about 1.5 billion years in the case of ^{40}K (after Press and Siever 1986, figure 2.28). Scale along vertical axis is arithmetic in (A) and logarithmic in (B).

nium, but totally rejects lead during its crystallization, reflecting size and charge requirements of cation sites in the structure. Uranium has two radioactive isotopes, ^{238}U and ^{235}U, which decay ultimately to isotopes of lead, ^{206}Pb and ^{207}Pb, respectively, after intermediate fission steps (involving, among other products, alpha particles—helium atoms). The rates of these decay reactions have been measured very accurately. Again, analyzing the proportions among the radioactive parental uranium, and radiogenic-lead daughter isotopes allows quantitative assessment of the age at which the host mineral formed (assuming the initial absence of lead in the host zircon and no import or export of atoms). Employing these and several other isotopic systems for a variety of phases, earth scientists are able to determine quantitatively mineral or bulk-rock radiometric ages in favorable circumstances. Strictly speaking, this apparent age is the time at which the mineral or rock in question became a closed system; in other words, when it ceased equilibrating isotopically with its surroundings. Some earth scientists refer to isotopic ages as absolute ages (quantitative), in contradistinction to fossil ages, which are all relative (qualitative).

Radiometric methods have allowed the assignment of times of formation to primitive chondritic (granular) bleb- or drop-containing meteorites, and to anorthosite breccias of the light-colored, low-specific-gravity lunar highlands, which are 4.55 and slightly in excess of 4.4 billion years old, respectively. These materials, which are substantially older than the dark lunar maria (basalt-floored lunar lowlands, thought by the ancients to represent seas), have not been appreciably reheated since condensation of the solar nebula—the time of origin of our solar system. Because they have not been thermally annealed since their formations, primitive chondrites and lunar highland breccias testify to the original collapse of the interstellar gas cloud, and origin of the Sun, as well as subsequent early accretion and differentiation of the Earth, the Moon, and the other planets.

Long before radioactivity was discovered, however, geologists had wrestled valiantly with the problem of the antiquity of the Earth. A basic tenet of stratigraphy (the study of layered—chiefly sedimentary—rocks formed at or near the surface of the lithosphere) is that, in structurally undisturbed sections, the oldest layer occurs at the base, with successively younger strata overlying this unit, and the youngest stratum resting on top. This law of superposition enunciates what a moment's reflection will corroborate: the oldest sediments in a section are laid down as a blanket resting on the basement of a depositional basin, with progressively younger strata deposited sequentially above. Similarly, the oldest lava bed occupies the basal portion of a stack of flows, with the most recent eruptive at the top. Thus the relative ages of such surficially deposited units can be determined readily by recognizing which way is up in the lithologic succession.

Two centuries ago, the geologic processes resulting in accumulations of layered sediments in lakes, rivers, and along the seashore were observed and the depositional rates roughly measured. It was found that sedimentary sequences apparently require hundreds of thousands—even millions—of years for the buildup of a few hundred meters of section. The great antiquity of the Earth was thereby inferred by postulating that current processes and rates must have characterized all of geologic time. This is the so-called hypothesis of uniformitarianism, an implicit affirmation that "the present is the key to the past." What it means is that the type of geologic activity taking place today (not just obeyance to the relevant principles of mathematics, and the laws of physical and biological sciences) also occurred in the distant past in the same fashion. As we will see later in this chapter, some geologic phenomena may have had a rather different aspect attending early stages in the development of the Earth.

Support for the law of superposition has been provided by the nature of fossil materials entombed in strata, for it became clear early in the nineteenth century that the more ancient rocks contain strange, relatively primitive life forms (or none at all) compared with the progressively more familiar, organizationally more complex animal and plant remains contained in the overlying, younger beds. Indeed, these marked contrasts of distinctive fossil assemblages from different geologic ages are ubiquitous on Earth, and have allowed the time correlation of geographically remote stratified rock series and the assignment of relative ages of formation. Darwin's theory of evolution provided a mechanism to explain the observed changes in life through the geologic ages, and supported the concept of uniformitarianism.

However, prior to the introduction of the radioactive dating method, geologists had only semiquantitative extrapolation techniques, such as that based on observed sedimentary accumulation rates, with which to gauge the numerical—or absolute—ages of rocks. Moreover, macroscopic fossils are largely absent from strata laid down prior to the Cambrian Period, about 570 million years ago, so only crude relative ages could be assigned to lithic units older than the Phanerozoic Era, or time of megascopic life. These earlier, Precambrian times have been termed the Proterozoic Era or time of primitive life, and the yet older Archeozoic Era, or time of ancient life (now generally referred to as the Archean). The "lost" interval between the formation of the Earth and preservation of the oldest rocks (4.5–3.8 billion years ago) is called Hadean time, in poetic reference to the inferred, especially hot, turbulent conditions of accretion of the planet.

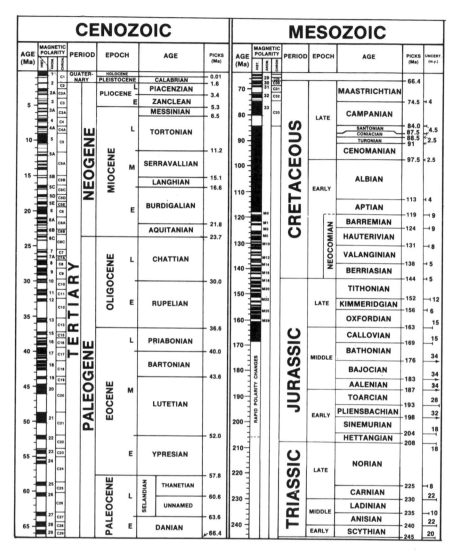

FIGURE 7.2. The geologic time scale, (modified from the Decade of North American Geology, Geologic Society of America, 1983). The Cenozoic, Mesozoic, and Paleozoic eras in aggregate constitute the Phanerozoic, or time of macroscopic

The parade of life through the ages is an exciting, engrossing topic in itself, but is not treated in this book. Here we will simply present the geologic time scale (figure 7.2). Note that the time of megascopic life represents only about the last eighth of the interval that has lapsed since

PALEOZOIC

AGE (Ma)	PERIOD	EPOCH	AGE	PICKS (Ma)	UNCERT. (m.y.)
260, 280	PERMIAN	LATE	TATARIAN	245	20
			KAZANIAN	253	20
			UFIMIAN	258	24
			KUNGURIAN	263	22
		EARLY	ARTINSKIAN	268	12
			SAKMARIAN		
			ASSELIAN		
300, 320	CARBONIFEROUS / PENNSYLVANIAN	LATE	GZELIAN	286	12
			KASIMOVIAN	296	10
			MOSCOVIAN		
			BASHKIRIAN		
				315	20
340, 360	CARBONIFEROUS / MISSISSIPPIAN	EARLY	SERPUKHOVIAN	320	
				333	22
			VISEAN		
				352	8
			TOURNAISIAN	360	10
380, 400	DEVONIAN	LATE	FAMENNIAN	367	12
			FRASNIAN	374	18
		MIDDLE	GIVETIAN	380	18
			EIFELIAN	387	28
		EARLY	EMSIAN	394	22
			SIEGENIAN	401	18
			GEDINNIAN	408	12
420	SILURIAN	LATE	PRIDOLIAN	414	12
			LUDLOVIAN	421	12
		EARLY	WENLOCKIAN	428	8
440			LLANDOVERIAN	438	12
460	ORDOVICIAN	LATE	ASHGILLIAN	448	12
			CARADOCIAN	458	16
		MIDDLE	LLANDEILAN	468	16
480			LLANVIRNIAN	478	16
		EARLY	ARENIGIAN	488	20
500			TREMADOCIAN		
				505	32
520	CAMBRIAN	LATE	TREMPEALEAUAN		
			FRANCONIAN		
			DRESBACHIAN	523	36
		MIDDLE			
540				540	28
560		EARLY			
				570	

PRECAMBRIAN

AGE (Ma)	EON	ERA	BDY. AGES (Ma)
750	PROTEROZOIC	LATE	570
1000			900
1250		MIDDLE	
1500			1600
1750, 2000		EARLY	
2250			
2500			2500
2750	ARCHEAN	LATE	
3000			3000
3250		MIDDLE	3400
3500		EARLY	
3750			3800?

life. The Precambrian consists of the Proterozoic or time of primitive life, and the Archeozoic (Archean), or time of ancient life.

the formation of the Earth. Primitive algae were extant 3.5 billion years ago, at the latest, so most of the Precambrian terrestrial history was characterized by infinitesimally slowly evolving primitive life forms prior to the spectacular flourishing of more advanced plants and animals during the Phanerozoic.

⸗ PHYSICAL EVOLUTION OF THE HADEAN EARTH

As we saw in chapter 1, presuming that the Earth accreted as an initially homogeneous body, the interior of the planet (mantle plus core) must have undergone profound thermal, mineralogic, and chemical changes over the course of geologic time. Direct evidence concerning these primordial events is limited by an inability to sample and measure intact portions of even the uppermost mantle. The atmosphere and hydrosphere also have evolved with time: because of diffusion of light, low atomic number, volatile species into outer space; and through chemical interactions of the global fluid and gassy envelopes of the Earth with the crust and, indirectly, the upper mantle. However, circulation and mixing of the atmosphere and hydrosphere, extremely rapid on a geologic time scale, have thoroughly destroyed evidence of the primordial stages. Accordingly we examine the rocky outer shell of the Earth for clues regarding the earliest phases of development, because the surviving crust contains the only accessible record of the differentiation of the primitive planet. Modern and probably also ancient styles of sea-floor spreading and subduction have resulted in the recycling of the oceanic crust and its underpinnings (at current rates, about once every 100 million years); hence, information regarding the nature of the early Earth resides almost exclusively in the continents.

The sialic crust, however, represents only a very small portion of our chemically differentiated planet. To understand the production of this near-surface "slag," we must briefly review the inferred history of the core and mantle. Differentiation of the Earth as a whole, of course, can be placed in better perspective through consideration of the origin and general evolution of the inner, terrestrial planets. Besides this, the Hadean rock record is well preserved on the Moon, and apparently, on most of the other inner planets except for the Earth. Therefore, we will briefly speculate on the formation of the Earth, in relation to the Moon, Mars, Mercury, and Venus, building on the planetary and meteoritical data presented in chapter 1.

Because of the relative abundance of the long-lived primordial radioactive isotopes of elements such as uranium, potassium, and thorium in the newly formed Earth—and the initial presence of short-lived radioisotopes as well—a significant amount of planetary self-heating would have occurred during and directly following condensation of the solar nebula. Moreover, most planetary materials except for metals are poor thermal conductors; radioactive and accretionary (impact) heating thus would have promoted incipient melting of the outer portion of smaller condensed planetary bodies and a substantial degree of partial fusion for the more massive ones such as the Earth and Venus. Because iron melts at a lower temperature than do silicates under high confining pressure, and because silicate and liquid iron (plus nickel) are immiscible, dispersed droplets of dense melt rich in iron (plus nickel) must have formed at depth within the Earth. Due to gravitational instability, these droplets would have migrated downwards, coalescing in the process and displacing the silicates upwards. The end result would have been a molten metallic core overlain by a largely solid, silicate mantle (see figure 1.10). The conversion of kinetic energy to heat during this infall, as it is called, would have liberated substantial additional amounts of thermal energy, contributing to the fusion and more complete separation of molten metal core and refractory silicate mantle, and partial melting of the latter. Slight cooling since this earliest stage of planetary differentiation has resulted in formation of the Earth's solid inner core. Differential circulation in the outer, liquid part of the core provides the dynamo that is probably responsible for the Earth's enveloping magnetic field. By analogy with the Moon, separation of the iron (plus nickel) core and overlying silicate mantle, as well as generation of a partly molten upper mantle, or magma ocean, was completed no later than about 4.4 billion years ago.

No terrestrial rocks have survived from this early, Hadean stage of planetary evolution; indeed, the most ancient samples that we now know, from northern Canada and western Greenland, are "only" about 3.9 and 3.8 billion years old. However, evidence of an analogous initial stage of planetary development is well preserved on the Moon. The plagioclase feldspar–rich impact breccias of the lunar highlands (rocks pulverized by meteoritic bombardment) contain fragments at least as old as 4.4 billion years; the anorthositic highlands are heavily cratered (figure 1.4B), attesting to continued, intense planetesimal influx there— and undoubtedly on Mercury, Venus, Earth, and Mars—to about 3.9 billion years ago. Meteoritic accretion probably was most intense during the earliest stages of planetary formation and tapered off toward the end of Hadean time.

At least parts of the older highlands appear to have formed from the gradual accumulation of calcic plagioclase that rose toward the surface of the lunar magma ocean 4.4–4.5 billion years ago. The lunar maria are floored by Early and Middle Archean basalts; evidently the thermal budget of the Moon allowed local production of mafic magma within its mantle, followed by upwelling and extrusion as late as 3.9 to 3.0 billion years ago. Older maria are heavily pockmarked, whereas the younger maria basalt fields are much less scarred by impact craters; this relationship supports the idea that meteorite bombardment tailed off dramatically during the Early Archean.

Because of the far greater mass and gravitational attraction of the Earth, higher-pressure equilibria would have dominated terrestrial crystallization other than in the most surficial parts of an analogous, now obliterated terrestrial magma ocean. Accordingly, on the Earth, early crystallization of dense minerals such as garnet (specific gravity of 3.6) from the partly molten mantle under attendant high pressures (rather than the corresponding calcic plagioclase (specific gravity of 2.7) for the Moon) would have resulted in the settling of aluminous crystalline phases rather than their flotation. In addition, on Earth, the thin, transitory, refractory crust that would have solidified at the upper cooling surface, lying upon less dense molten material, would have foundered (sunk) as a consequence of perturbation by meteoritic bombardment. During crustal production, therefore, dense, refractory ferromagnesian minerals should have been transported to, and stabilized in, relatively deeper portions of the gradually thickening, largely solid mantle of the primitive Earth; concomitantly, the more fusible and volatile elements would have been concentrated toward the surface as a relatively silicic, globe-encircling rind and dense primitive atmosphere. Until temperatures fell below those at which granite liquifies (melting begins at temperatures of 600–1000 degrees Celsius, depending on the abundance and pressure of H_2O), this more silicic, alkalic material would have remained largely molten, in contrast to the more refractory calcium-rich plagioclase accumulations. Probably during this early, very hot, Hadean stage, the ephemeral sialic material would continue to have been reincorporated in the rapidly convecting, rehomogenizing mantle due to viscous drag. Outgassing would have been intense because of the high temperatures and very active volcanism, with the volatiles transferred from the solid Earth to the dense early atmosphere.

By 3.8 billion years ago, near-surface conditions had ameliorated considerably; water-laid sediments containing silicic, fusible, continental crust-type debris were being deposited on our planet. Clearly, the surface temperature was less than the boiling point of H_2O. Moreover, sialic crust, from which the sediments were derived, was present. Thus, layers of continental plus oceanic crust and enveloping hydrosphere plus atmosphere must have existed during Late Hadean time. The partial fusion of mantle peridotite requires higher temperatures (on the order of 1200 degrees Celsius) than necessary for formation of continental crust; thus at least a thin, solid upper mantle layer must also have been present by this stage (see figure 7.3). We may safely conclude that the chief petrologic components of the lithosphere, as well as both primitive hydrosphere and atmosphere, existed on Earth before or, at the latest, by the beginning of Archean time.

This condition succeeded, and must have overlapped the earlier,

Hadean stage of Earth history. By analogy with the Moon, this precursor stage probably was characterized by intense meteoritic bombardment and planetesimal accumulation during fractional crystallization of the terrestrial magma ocean, widespread, exuberant volcanism, abundant release of volatile constituents, sialic recycling through crustal foundering, and meteoritic reworking.

The phase diagram for the systems granite plus fluid (equal continental crust) and peridotite plus fluid (equals mantle) is presented as figure 7.3. Melting relations are shown both in the absence of volatile constituents and in the presence of minor or excess fluid. For lithological environments lacking a readily dissolvable volatile component such as H_2O, increased pressure elevates the temperature for the onset of melting. In contrast, the presence of abundant H_2O (and, at upper mantle pressures, CO_2), causes a lowering of rock fusion temperatures as pressure is raised. This is a consequence of the increasing solubility of these constituents in silicate melts with increasing P. Sialic material is less refractory than basaltic and peridotitic material, so the first continental crust must have formed on a solid substrate of oceanic crust and mantle. Hence the generation of sial must have begun on the early, hot Earth after the establishment of lithospheric segments (platelets) surmounting the volumetrically dominant asthenosphere.

Calculated geothermal gradients depend critically on the assumptions employed. Nevertheless, the early Earth unquestionably had an

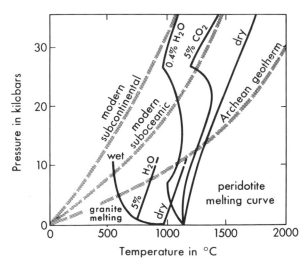

FIGURE 7.3. Phase relations for the initiation of melting of primitive peridotite, anhydrous and in the presence of minor amounts of H_2O and of CO_2, and of granite, dry and in the presence of minor and excess H_2O (from Ernst 1983). Also shown are computed present-day intraplate continental and oceanic geothermal gradients, and an estimated 3.6-billion-year-old averaged geotherm, assuming approximately three times the current radiogenic heat production and a compositionally layered Earth.

average temperature/depth curve higher than currently exists, as required by impact heating, the large terrestrial inventory of heat-producing radioactive elements, and by infall of the core (see following section). Continental crust and mantle would have begun to melt at shallower depths than at present because of the high heat generation, heat flow, and geothermal gradient. For this reason, Archean continents and their underpinnings of mantle lithosphere may have been, on the average, relatively thin compared with present-day relationships.

HEAT GENERATION AND ENERGY TRANSFER
IN THE EARLY EARTH

The chemical/density differentiation of a planet is largely a consequence of thermal energy transfer operating in a gravitational field. As we have seen, the elevated heat production provided by primordial abundances of radioactive elements, additional energy supplied by meteorite impacts, and rapid infall of the iron (plus nickel) core would have resulted in high temperatures throughout Hadean time. The computed heat flow contributed solely by radioactive decay employing two differing assumptions is shown in figure 7.4. In addition, energy released during rapid core formation would have provided enough heat to raise

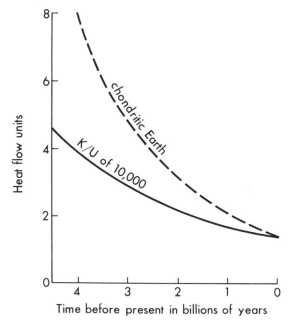

FIGURE 7.4. Variation of heat flow (in microcalories per cm^2 per second) reaching the surface due to radioactivity as a function of geologic time (summarized by Ernst 1983). The dashed curve is for the assumption of a primitive chondritic meteorite chemistry for the Earth; the solid curve is appropriate for a potassium/uranium ratio of ten thousand to one (somewhat lower than undepleted chondrite). Both models assume whole-mantle convection since Hadean time; the more likely layered circulation would provide a more complicated heat flow pattern.

the Earth's temperature more than 2000 degrees Celsius. Even with rapid mantle convection and heat liberation, it is difficult to imagine how such a thermal event as core formation could have avoided producing a near-surface Early Hadean magma ocean. The segregation of the core certainly required fusion of the metal, and the present-day solid inner core is testimony to the fact that this central region has cooled slightly over the course of geologic time. Apparently so has the outer Earth, for by 3.8 billion years ago at the latest, liquid water was present and was responsible for the sedimentary reworking of pre-existing continental crust.

The planet has redistributed heat by several thermal transfer mechanisms, notably through radiation, conduction, and mass flow (convection). Convection appears to be the most efficient process for moving thermal energy from the deep interior toward the surface, and it may have operated since core formation in Hadean time. As discussed in chapter 2, present-day motions of lithospheric plates are generally regarded as a manifestation of bodily circulation within the mantle. The convection reflects a thermally driven gravitative instability, a result of the dense, cool lithosphere overlying less dense, hot asthenosphere. Circulation and overturn of the mantle occur whether due primarily to buoyant thermal anomalies at depth (plumes), or to slip of the dense lithosphere off topographic prominences (ridges) and/or down subduction zones (trenches).

Currently, most of the terrestrial heat loss occurs in the ocean basins, especially in the general vicinity of spreading centers. These ridge systems are the uppermost part of the hot, rising limbs of mantle convection cells. In contrast, stable continents are characterized by much lower thermal fluxes. Descending lithospheric slabs are heat sinks for the mantle, so heat flow measured near trenches is abnormally low. The early Earth probably was not more homogeneous in terms of heat flow; if anything, the much higher energy contents and fluxes would have required more marked disparities between regimes of high thermal dissipation compared with low heat-flow zones. Ancient igneous and metamorphic rock types provide important clues as to the nature of these contrasting regimes. Ultramafic lavas (komatiites) of Archean greenstone belts require temperatures exceeding 1600 degrees Celsius for their production and reflect moderate to large-scale melting of upwelling mantle. Such substantial degrees of melting could only have been generated in oceanic regions of very high temperatures and heat flow. Thick Archean crustal sections of high-grade sialic gneiss, on the other hand, contain phase assemblages suggestive of less extreme metamorphic geothermal gradients, indicating that these portions of the continents were not subjected to especially elevated thermal fluxes.

Small, rapidly convecting upper mantle asthenospheric cells apparently would have allowed the efficient transfer of larger quantities of heat from the mantle toward the surface and ultimately out into space by radiation. Thus, in contrast to the modern lithospheric plates of great thickness and lateral extent, driven by large-scale, stately upper mantle circulation, the early Earth probably was characterized by oceanic areas underlain by small, rapidly overturning mantle convection cells, with continental debris accumulating in cooler eddies above regions of asthenospheric stagnation or downwelling. Organized flow and overturn in deeper, denser portions of the lower mantle since core formation is likely, and would mean that, since early in its history, the Earth's mantle has been characterized by tiered layers of convection cells.

⎯ GENERATION OF THE CONTINENTAL CRUST

The ocean basins contain crust as old as about 190 million years as a maximum, and of course, most of it is much younger. This basaltic/gabbroic material is produced at spreading centers by partial fusion accompanying the rise of somewhat depleted mantle material (that is, primordial chondritic material minus a small fraction of earlier generated basalt); the process results in a further depletion of the ascending peridotite (chapter 4). This uppermost part of the oceanic lithosphere is recrystallized, hydrated in part, and continuously returned to the mantle today. It was probably similarly reworked during the early stages of planetary evolution. The fate of this lithospheric plate complex on descent into the present mantle is a matter of debate, but evidently its devolatilization and partial melting produces andesitic to dacitic magmas as well as basalts characteristic of active continental margins and island arcs (chapter 5). Intermediate and silicic materials are less refractory (figure 7.3) and, more importantly, are less dense than subjacent portions of the mantle lithosphere. Hence, they tend to decouple from downgoing plates and remain near or at the surface due to buoyancy. Sial, therefore, has accumulated over time, along with scraps of oceanic crust resulting from accidental tectonic insertions. This continental crust is a geologic collage constituting the most important source of information concerning stages in the development of the Earth.

However, in Hadean time, mantle circulation apparently was so vigorous that virtually all of the sialic material produced is conjectured to have been returned to the mantle because of viscous coupling. Isotopic evidence indicates that complete rehomogenization continued until about 3.6–3.8 billion years ago, that the average age of the continental crust is approximately 2.3 plus or minus 0.5 billion years old, and that much of the Earth's sialic material—perhaps 70 percent or more—had been generated by Early Proterozoic time. Thus, net preservation of sial was nil until the Early Archean, due to eventual reincorporation back into the mantle. Apparently about 3.6–3.8 billion years ago, the rate of return to the mantle gradually fell below the rate of formation of continental crust, and increasing amounts of sial accumulated at and near the surface, reflecting high—but decelerating—rates of mantle convection. This process gradually depleted the upper mantle in low-melting constituents and, as energy input to the thermally driven gravitative instability waned, the rate of production of both oceanic and continental crust from progressively more depleted mantle declined toward the modern, low generation rates. As portrayed in the schematic diagram of figure 7.5, the net accumulation of sialic material seems to have risen to a maximum in the Late Archean and Early Proterozoic and has gradually lessened ever since.

Mantle-derived igneous rocks constitute the primary additions to the Earth's crust. Properly sampled, such materials testify to the nature of this growth. Sedimentary processes rework the crust and, in most cases, result in chemically disparate units whose aggregate bulk composition, including the associated volatile species, reflects the chemistry of the source. The presently existing proportions and character of surficial layered rocks preserved from earlier stages of crustal development are

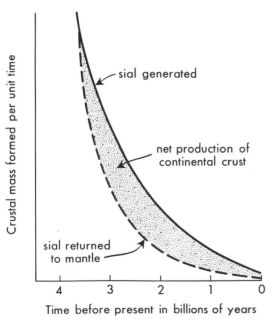

FIGURE 7.5. Diagrammatic representation of the competing processes of production (derivation from the mantle) and consumption (return to the mantle) of sialic material during mantle differentiation, and of net accumulation of continental crust (Ernst 1983). The total mass of the crust (shaded) is represented in arbitrary units.

illustrated in figure 7.6. The relative abundance of volcanogenic units evidently has decreased steadily over the course of geologic time, reflecting both the Earth's declining heat budget and the progressive chemical differentiation of the upper mantle. As will be discussed in the following sections, immature clastic sediments and banded iron formations (chemically precipitated ferruginous—i.e., iron-containing—silicia-rich sediments) are volumetrically important in Late Archean and Early Proterozoic units, whereas younger deposits are typified by a gradually increasing proportion of multicycle, chemically differentiated, platform strata.

—— LITHOLOGIC ASSEMBLAGES THROUGH GEOLOGIC TIME

The Archean Rock Record

Early Precambrian lithotectonic entities may be divided into two principal petrologic groups, granite plus greenstone belts and high-grade metamorphic gneiss complexes. The former contain abundant associated metasedimentary sequences, especially in their upper stratigraphic levels, and in aggregate display evidence of predominantly vertical dif-

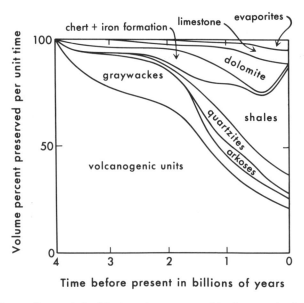

FIGURE 7.6. Proportions of stratified rocks preserved in the geologic record per unit time (after Condie 1982).

ferential movements; the latter typically contain low-angle thrust faults and overturned fold structures, indicative of subhorizontal differential compressive movements. Broad tracts of high-grade gneiss characterize portions of the western Greenland–Labrador, south Indian, and Antarctic cratons, whereas feebly recrystallized greenstone plus granite assemblics typify portions of the Canadian, western Australian, and South African cratons (cratons, or shields, you will remember are low-standing, worn-down stable platforms, constituting the ancient Precambrian cores of the continents). In general, however, high-grade and low-grade metamorphic complexes are associated with one another on a regional scale.

Some of the oldest preserved rocks on Earth crop out as a high-grade gneiss terrane in western Greenland; they consist of several types of amphibolite, ultramafic lenses, metacherts, carbonate-bearing mica schists, and metaconglomerates that contain volcanic clasts. This stratified sequence has been engulfed by 3.7-billion-year-old medium- to high-grade granodioritic gneisses, which undoubtedly were igneous plutons prior to a later stage of dynamothermal metamorphism. A generalized geologic map of these high-grade Early Archean rock units from western Greenland is illustrated in figure 7.7. Granites and relatively high-grade feldspathic gneisses of nearly comparably age crop out in southwestern Minnesota, in southern Africa, in eastern Antarctica, in northern Canada, and in eastern India.

Archean complexes exposed in South Africa, Zimbabwe, western Australia, peninsular India, Canada, Brazil, and Fennoscandia (Finland, Sweden, Norway, Denmark) are typified by belts of greenstone and superjacent volcanogenic sedimentary strata (largely immature, first-cycle sandstones) as old as 2.7–3.6 billion years. In general, these relatively small, linear complexes occur as weakly metamorphised, downward-bowed folds (synclinoria), surrounded by higher-grade feldspathic gneisses, migmatites, and pyroxene-bearing amphibolites. The down-draped nature of greenstone belts reflects postdepositional sagging that accompanied relative decent of the negatively buoyant mafic lavas and rise of the adjacent felsic granitoids. The downwarped structure of the greenstone belts, therefore, is a deformatiomal feature, not an original basinal configuration. The associate sodium-rich, potassium-poor gneisses are crudely upward-bowed, bulbous structures (anticlinoria). Their crosscutting relationships to the more feebly metamorphosed greenstone keels (downward projecting belts), as well as their higher metamorphic grades, are probably indicative of buoyant plastic flow attending ascent from deeper crustal levels. The ferromagnesian greenstone plus metasedimentary linear complexes consist of denser units compared with the quartzofeldspathic gneisses. Thus such differential movement would be expected to take place where this pre-existing density inversion was perturbed by either a tectonic event or by thermal softening, or both.

Figure 7.8 presents an example of a low-metamorphic-grade greenstone belt from the South African shield. Generally lacking from this belt are mature platform-type strata such as quartz-rich sandstones, thick layers of limestone, and widespread evaporitic sequences (salt deposits); uncommon, too, are highly aluminous shales, red beds (sandstone/shale sedimentary strata), carbonates, and banded iron formations of great lateral extent. Structurally, greenstone complexes seem to have been dominated by vertical tectonics. Overturned folds and low-angle thrust faults are present in Precambrian complexes but appear to be relatively rare in greenstone plus granite assemblies, as illustrated in the schematic cross section of figure 7.9. The igneous rocks prominent in Archean greenstone belts consist chiefly of pillow basalts, indicative of submarine extrusion; in some cases, these grade upward into andesitic lavas and pyroclastics. Such volcanic cycles are generally capped by

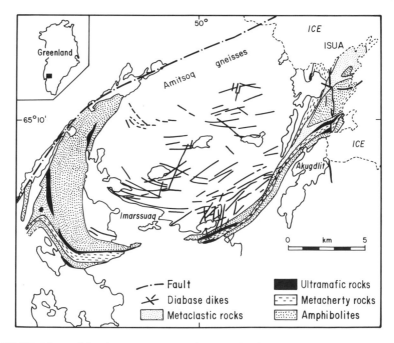

FIGURE 7.7. Map of the Isua supracrustal belt, a high-grade metamorphic complex from western Greenland (simplified after Allaart 1976). These approximately 3.8-billion-year-old rocks formed in a near-surface environment in the presence of seawater, as indicated by the occurrence of metamorphosed clastic and siliceous sediments.

immature clastic sediments (feldspathic conglomerates and sandstones), chert, or banded iron formation.

Komatiite and basaltic komatiite lava flows are the most distinctive rock types in the lower portions of many Archean greenstone belts. These ultramafic lavas are characterized by peridotite-like high MgO and low SiO$_2$ contents (see chapter 4). They must have been completely molten as demonstrated by the occurrence of intermeshed bladelike sheaves of prismatic olivine and orthopyroxene, typically a few centimeters long but, in rare instances, ranging up to a meter in length. These fragile bladelike crystals grew perpendicular to the surface of highly fluid, magnesium-rich flows, and could only have been produced after the lavas came to rest and congealed. The occurrence is significant, inasmuch as refractory komatiitic melts can exist only at temperatures exceeding 1600 degrees Celsius, even at very low pressures. Because this

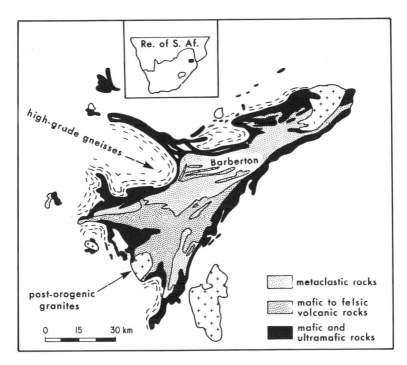

FIGURE 7.8. Map of the Barberton supracrustal belt, a low-grade metamorphic greenstone-plus-metasediment-plus-granite complex from South Africa (simplified after Anhaeusser 1971; and Viljoen and Viljoen 1971). Post-orogenic granites were intruded after folding produced the synclinal structures. At least some of the upper volcanics, and the sediments appear to have accumulated in a primitive island-arc environment.

rock type approaches essentially undepleted, fertile (primitive) mantle in composition, the generation of komatiitic liquid appears to reflect a high degree of partial fusion, hence elevated mantle temperatures.

Archean metasedimentary associates of the dominantly ultramafic, mafic, and more felsic igneous rocks of greenstone plus granite belts are commonly interpreted as representing poorly sorted, first-cycle clastics; these units are rich in volcanogenic debris and interlayered siliceous precipitates. Local derivation from a nearby volcanic chain and deposition at moderate to great water depths seem to be indicated. These belts contain no evidence for the existence of extensive tracts of nearby continental crust.

More obscure is the origin of those high-grade gneisses rich in sodium, potassium, and silicon that constitute major portions of the Archean cratons. Most have isotopic, major element, and trace element concentrations indicating a mantlelike source, but possess bulk-rock compositions typical of primitive continental crust. They evidently represent primary melts, or chemically immature, first-cycle clastic sediments derived from the solidification products of such magmas. These gneiss complexes may represent ancient, deeply buried, now extensively eroded analogues of dacitic-granodioritic batholithic belts such as the Andes, that occur at present-day convergent continental margins (chap-

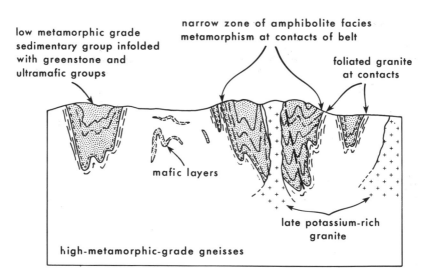

FIGURE 7.9. Diagrammatic cross section of low-grade greenstone synclinoria enveloped in a high-grade granodioritic gneiss complex (after Anhaeusser et al. 1969). Differential sinking of dense supracrustal units and rise of partly melted, deeply buried but buoyant quartzofeldspathic massifs are apparent.

ter 5). Considerable crustal thicknesses are required by the mineral assemblages of some Archean granulitic gneisses. For the western Greenland section, for eastern Antarctica, and for a number of other gneiss occurrences in the ancient shields, lithostatic pressures on the order of seven to twelve kilobars have been computed. This means that the Archean continental crust, at least locally, was twenty-five to forty kilometers thick. High heat flow and consequent high geothermal gradients would have resulted in wholesale melting and buoyant rise of the deeper portions of these sialic complexes. That they did not melt during metamorphism is clear from their preserved mineralogy and quartzofeldspathic compositions. The solid-state stability of surviving Archean gneisses and related rocks may be accounted for by the hypothesis that they evolved in relatively dry regions of the deep crust characterized by only modest heat flow. If a fluid phase attended recrystallization, it must have been CO_2-rich rather than H_2O-rich; otherwise, partial fusion and upward transport of buoyant H_2O-rich granitic liquid would have occurred. Indeed, fluid inclusions trapped in minerals from deep-seated granulites consist almost exclusively of carbon dioxide.

Archean Crustal Conditions

The occurrence in Archean sediments of water-worn sulfide grains, and of uraninite (uranium oxide) pebbles—minerals that are sufficiently resistant to survive long-distance transport and attendant dissolution, but only under strongly reducing conditions—suggests the presence of an early atmosphere/hydrosphere essentially devoid of free oxygen. The existence of discontinuous, thin sedimentary layers of stromatolitic banded iron formation in the Archean seems to be consistent with a reducing atmosphere apparently required for transport of soluble ferrous iron in aqueous solution across sedimentary basins and deposition in regions containing O_2-producing algal (stromatolitic) reefs, typified by slightly more oxidizing conditions. The ubiquity of anoxic shallow marine environments (quite low oxygen concentrations compared with present-day atmospheric values) was apparently maintained until about 2.0–2.5 billion years ago, permitting the accumulation of soluble Fe^{+2} in the oceans, then wholesale oxidation and precipitation of the great banded iron formations of South Africa, western Australia, Labrador, Minnesota/Ontario, and Brazil during Early Proterozoic time (see next section).

The virtual restriction of komatiitic lavas to the Archean Era suggests that, at least in the geologic environment in which greenstone belts were formed, early Earth geothermal gradients were considerably higher than today. The nearly total absence—or at least, lack of preservation—of blueschists, eclogites, and other high-pressure, low-temperature metamorphic rocks (see figure 4.19) from primitive subduction-zone belts also attests to an ancient elevated-heat-flow regime where these lithologies became incorporated in the continental crust. This is not surprising because, averaged over the entire Earth, heat production due to radioactive decay 3.6 billion years ago was more than three times the present rate, and was even greater during Hadean time (figure 7.4). The Archean abundance of pyroxene-bearing amphibolitic gneisses, and sillimanitic granulites likewise suggests relatively high crustal temperatures. High-grade gneisses are proportionately more abundant in Precambrian complexes than in Phanerozoic metamorphic sequences, presumably as a consequence of the ancient, moderately high-gradient thermal regimes. However, due to greater extents of erosion for the oldest terrestrial rocks, more deeply buried, high-T terranes are exposed, on the average, compared with younger orogenic belts; therefore part of the petrologic contrast in rocks of different ages may be due to sampling bias.

Lithologic evidence suggests that the early Earth, consisting of granite plus greenstone-plus-metasediment belts and higher-metamorphic-grade gneiss tracts, was characterized by the following factors: high lithospheric geothermal gradients; a variable, but in part, high degree of mantle partial melting; igneous rocks that were chiefly extrusive; an essentially anoxic atmosphere and ocean; and a veneer of locally derived, volcanogenic, and chemically immature sediments. The geothermal structure for the Archean Earth necessitated both by calculated heat flow and by the occurrence of preserved pyroxene-bearing gneisses and asthenosphere-derived komatiitic lithologic assemblages requires the presence of exclusively thin lithospheric plates and the generation of both oceanic- and continental-type protocrust (some thin, some thick) as relatively near-surface phenomena.

The extent and massiveness of this early crust is uncertain. Considering mineralogic data for the high-grade gneiss complexes, thicknesses of some Late Archean microcontinental massifs may have approached those of the present day. In contrast, based on igneous and metamorphic relationships displayed by the granite plus low grade greenstone plus metasediment belts, elevated geothermal gradients within these portions of the early Earth apparently promoted the partial fusion and upward rise of moderately deeply buried, H_2O-bearing sialic materials, resulting in the generation of predominantly thin crust. Two different heat-flow regimes, therefore, seem to be required to explain the contrasts between the profoundly buried, protocratonal high-grade gneisses on the one hand, and shallower-level circumoceanic granite plus greenstone plus metasediment belts on the other. The former apparently are characterized by relatively thick continental crust and a moderate geothermal gradient; the latter by thin, oceanic-island arc–type crust and a high geothermal gradient.

Lateral dimensions of the Archean plates are poorly constrained, but judging from the fact that most greenstone-plus-metasediment belts are relatively diminutive linear features, lithospheric plates and upper-mantle-flow regimes may have been considerably smaller than those of today. The existence of numerous little, rapidly circulating upper mantle convective cells would also be in accord with the high heat loss required by the inferred thermal structure of the early, rather hot Earth.

Because the pyroxene-bearing gneisses contain phase assemblages reflecting only slightly elevated continental and subcontinental geothermal gradients, most of the early heat dissipation must have taken place within a rapidly overturning suboceanic asthenosphere. Accordingly, oceanic lithosphere probably was thin, hot, and subductable chiefly due to viscous drag. Decompression partial fusion (i.e., partial melting of depressurizing solid material) of the rising suboceanic mantle convection limbs and/or plumes, characterized by moderately low volatile contents but high H_2O/CO_2 ratios, could have produced the observed, highly refractory komatiitic-basaltic suite typical of the greenstone belts. Many such mafic volcanic zones appear to have been formed in early Earth spreading-center regimes. Association of greenstones with overlying andesitic lavas and volcanogenic sediments may reflect chemical differentiation of the mafic liquid or proximity to a nearby island arc or active protocontinental margin. Most of the mafic and ultramafic materials were presumably returned continuously to the vigorously circulating, primitive mantle, but segments at the margins of oceanic cells might have become detached episodically and tectonically inserted in the cooler sialic environment. In the presence of an abundant aqueous volatile phase given off by a degassing, undepleted mantle, ultrametamorphic granitic plutons would have been mobilized near the base of the accreting continental or island-arc crust, and would have risen to upper levels, displacing the denser, overlying greenstone belts into downward-sinking folds accompanying thermal softening. Such oceanic crust and circumoceanic (sial plus sima) composite crust necessarily would have been thin because of the high geothermal gradient and elevated H_2O/CO_2 ratio of the associated fluids being expelled from the mantle (see figure 7.3).

The high-grade gneissic complexes represent thick accumulations of highly sodic, low-potassium granodioritic material; that which failed to melt must have been located in portions of the Earth well removed from the small, rapidly convecting, high thermal-flux suboceanic mantle cells. These accretionary regions could have been positioned over sites of stagnant or downwelling Archean asthenospheric flow. Even so, the metamorphic geothermal gradients inferred from the mineral assemblages would have promoted partial fusion of deep portions of the primitive continents unless the concentration of H_2O were low. Analyzed fluid inclusions trapped in minerals from such complexes are dominantly CO_2-rich, compatible with the persistence of thick, ancient crust only in an H_2O-impoverished environment.

But through what processes did these high-grade gneisses form? Mantle circulation resulted in the underflow of hot, soft, thin lithospheric slabs or platelets; shallow levels of partial fusion could have produced melts of the andesite-dacite series from pre-existing basalt, gabbro, and amphibolite. Or, the heating and incipient melting of thickened, predominantly mafic circumoceanic arcs in regions of moderate heat flow could have given rise to chemically more silicic and alkalic magmas, with the dense ultramafic residual solid assemblage settling back into the upper mantle. Both processes require the presence of aqueous fluids for the production of sialic suites, and would have been more or less confined to the relatively high-heat-flow circumoceanic regions. Once formed, however, thin segments of island-arc material could only have accumulated to great thickness in the relatively cool, devolatilized, or CO_2-rich environments of mantle downwelling discussed above.

Neither paleomagnetic (see next section) nor geochronologic data provide a basis on which to estimate quantitativley the area—or volume —ratio of Archean ocean basins to continents. The proportion of the Earth floored by oceanic, as opposed to continental, crust is a complex function of the volume and thickness of both, with the lateral extent and thickness of the continents being determined by a variety of factors. Among them: the geothermal gradient, and, thus, the depth at which partial melting of fusible, volatile-rich quartzofeldspathic material occurs; the volume of the hydrosphere, the freeboard (elevation above sea level) of the continents, and their potential erosion to wave base; the compositions of fluids being evolved during crustal growth; the rate of sweeping together of continental (island-arc) debris and partial return to the mantle via the mechanism of upper mantle circulation; and the rate of production of sialic material through chemical differentiation of relatively undepleted mantle. The Mohorovicic discontinuity must have been located at somewhat shallower depths during Early Precambrian time, especially in the circumoceanic realm, because of the relatively high geothermal gradient. The masses of both hydrosphere and continental crust apparently have increased—but at a lessening rate—over the course of geologic time, the result of continued, but waning, mantle evolution (see figure 7.5). The chemical differentiation of the mantle, in turn, is a consequence of the P-T gradient within the Earth.

The Proterozoic Rock Record

Available paleomagnetic data suggest the possibility that, with the notable exception of a proto–North Atlantic rifting/converging cycle during the early and middle parts of the Paleozoic Era, portions of Laurasia and two Gondwana supercontinents may have existed as discrete cratonal massifs for at least 0.8–1.6 billion years during most of the Proterozoic Era. This ancient continental assembly has been somewhat dispersed by the current cycle of continental drift, initiated during Late Triassic time. A diagram is presented in figure 7.10 illustrating the apparent positions, relative to the Canadian shield, of the north magnetic pole as a function of time (i.e., Early and Middle Proterozoic); this time-related change of pole positions determined through rock remanent-magnetism studies is called an apparent polar wander (APW) path.

With the exception of the 1.1-billion-year-old Grenville province, the various cratonal blocks appear to have moved as a single unit during the Early and Middle Proterozoic. About one billion years ago, the Grenville province was added to the continental collage, and thereafter it too drifted with the rest of the Canadian craton, judging from coincidence of APW paths. For rocks lithified during Early Proterozoic time, such apparent polar wander curves for South America and most of southern and western Africa approximately coincide; evidently at that time these sialic masses must have been connected to one another. A similar lateral continuity probably holds for the cratons of northern Africa and western Australia. Thus, broad continental assemblies were characteristic of the Proterozoic Era. Archean paleomagnetic data are fragmentary and much less definitive, however.

During the Early Precambrian, sialic island-arc materials seem to have been rapidly swept against one or more enlarging microcontinents, with other areas of the Earth's surface being typified by primitive oceanic crust underlain by vigorously convecting mantle. As indicated by the apparent coincidence of magnetic pole positions, relatively rigid continental crust–capped plates of greater lateral extent came into existence as early as about 2.6 billion years ago.

During Proterozoic times, intracratonal mountain belts formed across shield areas, as documented by abundant radiometric data, especially from Africa. Perhaps stagnant or downwelling mantle beneath the continental crust was becoming involved in larger-scale convection patterns.

Judging from the preserved stratified rock record, continental shelves and shallow seas of great breadth (which are a prerequisite for long sedimentary transport distances and consequent chemical differentiation) may not have existed until about 2.5 billion years ago. Archean

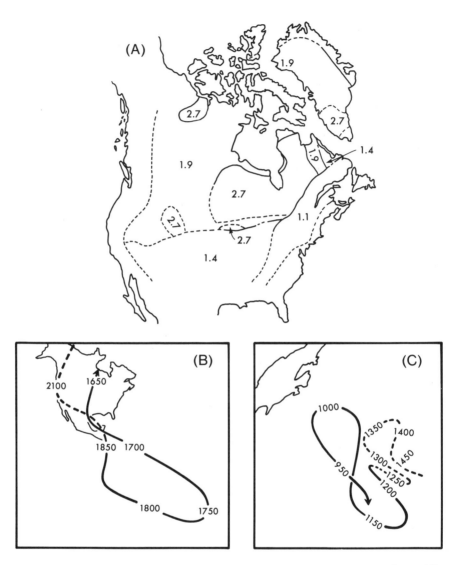

FIGURE 7.10. Portions of the Canadian shield (A), and apparent location of the north magnetic pole for (B) Early and (C) Middle Proterozoic times in millions of years (after Irving and McGlynn 1981). Segments of the craton contain much older rocks, but are differentiated on the basis of latest major mountain-building episodes, shown in billions of years: 2.7; 1.9; 1.4; and 1.1 (Grenville event). In (B) and (C), apparent polar wander paths are dashed where imperfectly known. The Grenville province accreted to the cratonal assembly by about 1.0 billion years ago.

sediments are typically first-cycle, locally derived, volcanogenic, and compositionally immature, whereas Proterozoic strata commonly include mechanically and chemically recycled units of much greater lateral continuity and compositional diversity (figure 7.6). The Archean microcontinental assemblies probably were small, scattered, and largely submerged, as suggested by the scarcity of platform sediments, with continental freeboard and widespread, shallow seas only being attained globally at the beginning of the Proterozoic Era.

The deposition of the thickest, most extensive banded iron formations occurred chiefly in the Early Proterozoic. This was evidently a time of transition between the Archean strongly reducing, and the more modern oxidizing, conditions in the atmosphere and shallow parts of the hydrosphere. Oxygen-producing organisms such as stromatolites (primitive colonial algae) were evidently responsible for this increase in oxygen concentration, a side-effect of which was the precipitation of ferric iron-bearing siliceous sediments. This biologically mediated change in the terrestrial atmosphere and hydrosphere toward more O_2-rich surficial conditions made possible the subsequent evolution of higher forms of life that require free oxygen in their metabolic processes.

The existence of continents, combined with transpolar drift allows for climatic extremes. Thus it is perhaps not a coincidence that most recognized ancient glacial deposits date from about 2.2 (possibly 2.7) billion years ago, shortly after the accretion of island arcs and microcontinental terranes had resulted in the construction of major continental assemblies.

Additional evidence supporting an increase from Archean to Proterozoic in the average thickness of sialic crust involves the manner of emplacement of basaltic liquids. Archean mafic igneous rocks are dominantly extrusive; large, floored gabboric intrusions appear in the geologic record only in units about 2.7 billion years old or younger. During the Archean, sial was repeatedly breached by fluid, ferromagnesian melts derived from depth, supporting the hypothesis that in many places the continental crust was relatively thin. By the end of this era, however, locally thick granitic crust apparently provided a gravitatively stable shield that, in some cases, could not be surmounted by dense mafic liquid. This meant that the hydrostatic head for such a magma was insufficient to permit its rise to the top of the relatively low-density, elevated continental crust, so intrusives spread out at depth instead of extruding and forming lava fields at the surface.

Clearly recognizable ocean-crust complexes date from the Proterozoic era (a few, much older ophiolites have been tentatively identified), and blueschists plus eclogites first appear in the geologic record in Late

Proterozoic terranes. Paired metamorphic belts (see figure 4.21) are virtually confined to the Phanerozoic. Therefore, attendance of the current style of lithotectonic processes, as described in chapters 2 and 4, can be evoked confidently only for the youngest Precambrian and more recent sections.

The Phanerozoic Rock Record

The modern style of plate tectonics has left a clear imprint on lithologic assemblages formed over the past 600–800 million years. Mountain-building processes and the areal disposition of igneous, metamorphic, and sedimentary provinces, while not understood completely, are familiar enough to require no further elaboration beyond material presented in chapters 2, 4, and 5. Probably the great linear extent of Phanerozoic mobile belts is at least in part a consequence of the large (inferred) lateral dimensions of present-day mantle convection cells (e.g., compare the modern volcanic belts of figure 2.16 with the seismically outlined plate junctions of figure 2.12). Indeed, it seems possible that the sizes and shapes of the capping lithospheric plates to a large degree determine the scale of lateral flow in the upper mantle.

The rifting and drift of continental crust–capped segments of lithosphere is a process that complicates Paleozoic and younger continental evolution. Although far-traveled (exotic) terranes evidently have played a major role in the reorganization of Phanerozoic crustal assemblies, the importance of rifting and drifting during the more substantial Precambrian growth of the continents is less clear. Archean accretion of granite plus greenstone plus metasediment belts to one another and to neighboring high-grade granulitic gneiss complexes represents the suturing of various presumably unrelated lithotectonic units, but the plate-tectonic settings and growth mechanisms of the individual terranes remain obscure.

Most described rock types of the Precambrian occur in Phanerozoic sections as well, but in differing volumetric proportions and structural environments. Conspicuously lacking from these more recent additions to the crust are komatiitic lavas and banded iron formations; uncommon, too, are high-grade gneiss complexes of great lateral extent. Perhaps equally significant, tracts of Archean rock in general lack multicycle, chemically differentiated platform strata (monomineralic carbonate units, quartz-rich sandstones, highly aluminous shales, salt deposits, red beds), glacial deposits, mafic layered intrusions, oceanic crustal associations, blueschists plus eclogites, paired metamorphic belts and kimberlites (these diamond-bearing pipes are thought to be derived from deep levels beneath stable cratons characterized by very thick lithosphere).

The Proterozoic Era was a time of transition; accordingly, rocks of this age exhibit a mixture of the early and late lithologic characteristics described above.

⸺ STAGES IN TERRESTRIAL PLATE-TECTONIC EVOLUTION

The geometries and present motions of lithospheric plates provide data regarding the mechanisms by which oceanic crust is formed and destroyed today. Modern growth of the continental crust seems to be due chiefly to the addition of andesitic and related rocks to the stable, nonsubducted side of convergent plate junctions. Effects of this process can be traced reliably back into sialic crust of Paleozoic and latest Precambrian age, where structural evidence of convergent plate boundaries exists. Sea-floor spreading, as it occurs today, produces a distinctive oceanic petrostratigraphic sequence that includes deformed mantle peridotites, overlying layered ultramafic-mafic complexes representing accumulated crystals from a magma chamber, diabase dikes and sills, and an overlying series of massive and pillowed basaltic lavas, breccias, and deep-sea sediments (see figure 2.6). Such ophiolite suites are found structurally thrust into Phanerozoic continental margins and island arcs, but few analogous units are recognized from the Proterozoic, and unambiguous examples of Archean ophiolites are very rare indeed. Paired metamorphic belts, similarly a typical product of Phanerozoic plate tectonics, appear to be absent from the Precambrian rock record. Of course, the farther back in time one searches, the more fragmentary and incomplete the preserved lithologic evidence becomes.

Terrestrial petrologic associations assignable to recognizable plate-tectonic settings, such as those that have characterized Phanerozoic times, seem to be lacking from all but the latest Precambrian. Is it possible, therefore, that plate-tectonic processes did not operate on the early Earth? The preservation to different degrees of heavily cratered surfaces on the Moon, Mercury, and Mars demonstrates that these bodies were not subjected to major crustal reworking through the subduction of lithosopheric plates subsequent to the marked decline in intensity of meteoritic bombardment approximately 3.9 billion years ago (see figure 1.2). Of course, all of these planets have considerably smaller masses than the Earth, and could not have long sustained core-formation, impact-, and radioactivity-induced internal temperatures as elevated as on the Earth; neither could they have maintained the strong thermal gradient required to fuel plate-tectonic processes for as extended a period. Although volcanic features demonstrate that primordial melting occurred on other terrestrial bodies, the Moon, Mercury, and Mars evidently cooled to the stage at which the thermally driven gravitative instability required for lithosphere-asthenosphere overturn died away prior to termination of the sweep-up of planetesimals. Moreover, some terrestrial (and planetary) volcanism does not appear to be related to mantle convection.

The larger masses of the Earth and Venus have allowed sustained retention of high temperatures and strong thermal gradients, the latter undoubtedly responsible for circulation within the mantle. Preliminary radar imagery from Venus, the Earth's sister planet, suggests that, although volcanic cones, rifted basins, and folded, layered rocks appear to be present, its topography is practically unimodal: Venus apparently does not possess substantially differentiated continents and ocean basins. Although plate tectonics may have characterized early stages of its development, such activity probably ceased on Venus before the end of Precambrian time. The virtual absence of H_2O there may be a consequence of devolatilization—or lack of condensation of hydrous minerals—due to proximity to the Sun, and a consequent run-away greenhouse effect, coupled with the gradual diffusional loss of H_2 to outer space. The lack of bimodal topography on Venus thus could reflect the inability of an anhydrous planet to manufacture sialic material without H_2O. The melting temperature of granitic, basaltic, and mantle constituents would have been substantially elevated in comparison with a wet environment (figure 7.3), thus precluding continental crust formation, and possibly even inhibition differential lithospheric plate motions on Venus.

Because bodily heat transfer has taken place since formation of the Earth's core in Early Hadean time, small, thin, soft platelets must have comprised the surface beginning quite early in the history of our planet, a supposition that is consistent with the preserved, if fragmentary, Archean rock record and with the phase relations illustrated in figure 7.3. Rapid mantle overturn probably would have drive such platelets against and beneath one another. Because of elevated near-surface temperatures and thinness of the plates, together with the small magnitude of the lithospheric/asthenospheric density inversion, the lithosphere would not have been subducted to profound depths in the asthenosphere before softening and losing its contrasting physical properties. This elevated thermal regime accounts for the almost complete lack of ancient high-pressure, relatively low-temperature metamorphic rocks such as blueschists and eclogites, as well as for the occurrence in Archean greenstone belts of the highly refractory komatiites and associated magnesian basalts. Andesitic and dacitic compositions could have been derived at shallow depths by fractional crystallization of mafic magma, through the partial fusion of thickened crustal sections of basaltic amphibolite, or by the incipient melting of hydrous upper mantle material.

Because of the high rate of outgassing and elevated heat flow of the primitive Earth, volcanic activity would have been much more volumi- nious than at present, resulting in the concomitant rapid early develop- ment of both mafic and quartzofeldspathic crust and the hydrosphere/ atmosphere. Archean high-grade gneisses were no doubt remobilized during accretion and thickening of the cratons; they possess mantlelike elemental and isotopic geochemistries, and could well have been derived by partial fusion from primordial peridotitic, basaltic, and amphibolitic precursors in a hydrous environment. Alternatively, they may represent metamorphosed and annealed thick accumulations of island arcs and microcontinental fragments, swept together by vigorous Archean sea- floor spreading. In any case, accumulation and preservation as thick Archean crust must have taken place above relatively cool, anhydrous portions of the mantle lithosphere and asthenosphere.

The Archean to Early Proterozoic transition seems to have been characterized by a gradual change from small, thin, hot Archean plate- lets, driven about frenetically by numerous, rapidly convecting asthen- ospheric cells, to relatively thicker, cooler, laterally more extensive, coherent lithospheric plates, the motions of which may have been a function partly of increasingly negative buoyancy and partly of ther- mally driven asthenospheric flow on a grander scale. At different times for different shields, but in general by the beginning of Proterozoic time, small sialic island-arc masses, produced chiefly during early mantle overturn and attendant chemical segregation, had become assembled into thick continental cratons. As freeboard thus became increasingly important, continental shelves became much more extensive, and me- chanical/chemical erosion and sedimentation began to play major roles in the production of chemically differentiated mature sedimentary facies.

As schematically illustrated in figure 7.11, four transitional stages in the plate-tectonic evolution of the Earth seem to be recognizable: (1) a Hadean tenuous lithosphere or preplate stage, characterized by a partly molten planet that was intensely bombarded by meteorites and that underwent profound gravitative differentiation to form a metallic iron (plus nickel) core, a ferromagnesian silicate mantle, and a continuously reworked ephemeral crustal scum; (2) an Archean platelet stage, in which the Earth's near-surface environment was dominated by hot, soft, rela- tively thin, nearly unsubductable platelets, and that aggregated sialic island-arc material to form protocontinents; (3) a Proterozoic supercra- tonal stage characterized by the emergence of broad continental shields, development of freeboard plus shallow seas, intracratonal mountain building, and the drift of supercontinents; and (4) a latest Precambrian- Phanerozoic cycle of modern-style plate tectonics involving rifting, the dispersal and suturing of continental fragments, subduction and the generation of long, linear, paired mobile belts at convergent plate mar-

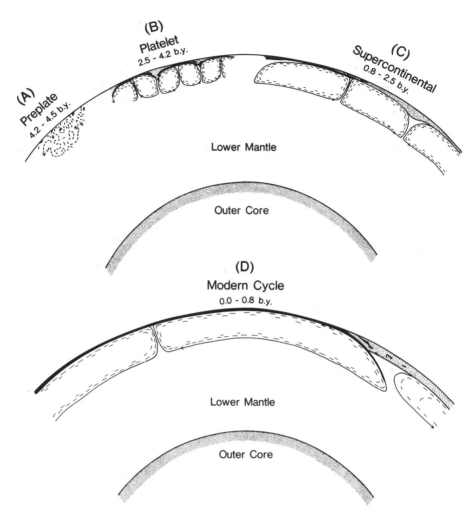

FIGURE 7.11. Sketch of hypothetical convective overturn in the Earth's upper mantle (and transition zone) as a function of geologic time. Ages are given in billions of years (abbreviated b.y.). Geometric relationships are diagrammatic and only approximately to scale. Oceanic crust–capped lithosphere (not just crust!) is shown in black, continental crust–capped lithosphere in stippled pattern. The Hadean stage (A) apparently was characterized by vigorous, ill-organized flow; the Archean (B) by more regular but rapidly convecting small cells and a few larger, stagnant regions; the Proterozoic (C) by laminar flowing cells of intermediate size and overturn rate; and the Phanerozoic (D) by giant, more slowly convecting, laminar-flowing cells. Depth to the base of the transition zone, and vertical thickness of the lithospheric plates, are presumed to have increased slightly over geologic time because of the waning geothermal gradient. Slowly moving or gently sinking regions of the upper mantle, thought to have been present during Archean time and to have been capped by enlarging granite-greenstone amalgams being annealed and transformed to high-grade gneiss terranes, are not illustrated in the Archean circulation pattern.

gins. Concomitantly, upper mantle convection cells were evolving through comparable stages—(A) chaotic to (B) small to (C) large to (D) enormous —with the ratio of lateral width to vertical thickness (aspect ratio) of the circulating mantle flow increasing over time.

The above scenario is obviously speculative. Indeed, because of the fragmentary and incomplete nature of the ancient rock record and the fact that studies of the tectonic evolution of the Precambrian Earth are still in their infancy, there are relatively few firm facts on which to base a sweeping, all encompassing overview. The lithosphere does seem to have thickened with the passage of time. In the Archean Earth, as today, two distinct crustal regimes were present, but the heat-flow and thickness contrasts may have been more pronounced then than now: the oceanic and circumoceanic (island-arc) regions characterized by relatively thin basaltic-komatiitic crust and by very high temperature/depth gradients and the protocratonal regions made up of andesitic-granodioritic crust and typified by moderate temperature/depth gradients. In spite of the uncertainties, it is important to recognize the changing character of the unidirectional petrotectonic processes that have affected the Earth's crust. This evolution, based on enlarging, ever more slowly circulating, convecting cells in the upper mantle, has been a response chiefly to chemical differentiation of the entire planet in the context of a gradually declining thermal budget.

☰ A FINAL WORD

We have briefly chronicled the origin and evolution of the Earth and its constituent materials from condensation of the solar nebula about 4.5–4.6 billion years ago to its present configuration. Buried heat due to radioactivity, infall of the core, and meteorite impacts, coupled with cooling at the terrestrial surface, have fueled circulation within the Earth's outer core, and in two or more layers of flowing cells in the largely solid mantle.

Since Hadean time, bodily flow at various depths has evidently been responsible for establishment of the terrestrial magnetic field, the efficient gravitative and geochemical differentiation of the planet, the contrasting motions of the several, evolving lithospheric plates, and the construction of on-land mountain ranges, submarine ridges plus trenches, and both oceanic and continental crusts. Drifting of the continents is yet another subtle but intellectually spectacular manifestation of the internal dynamic character of the Earth. The natures of minerals and rocks are manifestations of atomic forces on the one hand, and planetary thermal/chemical/dynamic mechanisms on the other. Mineral and energy resources as well as geologic hazards are intimately involved aspects of terrestrial evolution.

The web of life, that tenuous, self-replicating but ever-changing interaction of myriads of species among themselves and with the lithosphere, hydrosphere, and atmosphere, is vulnerable to the occurrence of geologic hazards as well as to the exceedingly rare but devastating arrival of impacting asteroids. On a continuing basis, all organisms of the biosphere are in one way or another critically dependent on the outer membrane of the lithosphere for sustenance. For this reason, man's accelerating utilization of natural resources, development and disruption of the physical and biological environment, and solid/liquid/gaseous pollution are alarming problems society must address immediately. Acid rain and global warming are but two of the looming environmental crises.

Sooner or later, we must reach equilibrium with our terrestrial substrate in order to survive as a species. Reexamine figure 6.4 in order to place matters in proper perspective. Tied to the increase of population, unlimited exponential growth of materials and energy usage, curve A, cannot be long sustained on a finite Earth. Our choices, therefore, are limited to curves B and C, renewable and exhaustible resource utilization, respectively. The rational program of development to follow is one that would allow world population to stabilize at a level such that usage of energy and earth materials would not exceed the rate of natural renewal. These environmental problems are urgent and must be attacked now. Solutions will not be easy or complete, but we dare not ignore the problems. Your generation will make the choices one way or another.

☰ GLOSSARY

Abundant Resource: a natural resource, the concentration of which in the accessible parts of the Earth is very high; accordingly, deposits of economic value must be very rich in the resource in question, and must be in significantly above-the-average terrestrial concentration. The oxides of silicon and iron are good examples of abundant resources.

Accretionary Wedge: a complex consisting largely of sedimentary units (including fragments of oceanic crust) scraped off, or decoupled from, the downgoing lithospheric plate near a convergent junction, and laminated beneath — shuffled up again — the stable, nonsubducted plate. Also called an accreting prism, or a subduction complex. (See figures 2.7, 4.10, and 4.21.) The shallow accretion is called offscraping, the deeper accretion, underplating.

Acid Rain: as gas species such as CO_2 and SO_3 dissolve in water, they form acids through reactions of the sort $CO_2 + H_2O = H_2CO_3$ (carbonic acid), and $SO_3 + H_2O = H_2SO_4$ (sulfuric acid). When these and other noxious gasses (NO_2, for instance) accumulate in the atmosphere due to the combustion of fossil fuels by industrial, vehicular, and home usage, rainfall becomes acid, polluting lakes and rivers.

Agglomerate: pyroclastic layered deposit containing volcanic bombs (see ejecta) set in a matrix of ash and/or lava.

Algal Reef: a biogenic carbonate bank (a bioherm) formed by the accumulation of $CaCO_3$ exoskeletons secreted by colonial algae.

Alpha (α) Particle: a radiation product that is actually a helium nucleus (two protons, two neutrons), bearing a charge of $+2$. It is produced by nuclear fission of a relatively high atomic number atom.

Amorphous: any solid or liquid that lacks a long-range ordered atomic arrangement—or crystal structure. Polymerization of atoms, ions, or molecules is therefore limited.

Amphibolite: a lineated, medium-grade metamorphic rock characterized by plagioclase intermediate between calcium and sodium, major amounts of hornblende and, typically, epidote and/or garnet. (See figure 4.19.)

Andesite: a volcanic rock—characteristically lava—intermediate in chemical composition between basalt and rhyolite. (See chapter 4 and table 4.1 for description and classification.)

Anhedral: adjective denoting a mineral grain bounded by irregular surfaces rather than by crystal faces.

Anion: a negatively charged ion ($Z < e$); so-called because, in an electrolyte solution, negatively charged ions migrate toward the positive electrical terminal, or anode.

Anion Complex: the covalent bonding of several anions about a central cation, which yields an overall negative charge to the structural subunit. Examples include CO_2^{-2}, SO_4^{-2}, and NO_3^{-1}.

Anomaly: a disparity from the anticipated relationship. In the earth sciences most commonly used for geophysical departures from a regional field gradient; e.g., gravitational, magnetic, seismic transmission velocity, thermal, etc.

Anorthosite: nearly monomineralic felsic igneous rock consisting almost exclusively of calcic plagioclase.

Anoxic: reducing conditions, severely depleted in oxygen.

Anticline: an archlike or domical fold in layered rocks in which the beds dip downwards away from a central high. (See figure 6.8.)

Anticlinorium: a great, linear, upwarp consisting of subparallel, dominant anticlines and subsidiary synclines. Basically a large anticline with superimposed minor synclines and anticlines (folding at two different scales).

Apparent Polar Wander Paths: presuming that a segment of the Earth's crust is fixed in its geographic location relative to the spin axis (the true north and south poles), the remnant magnetism of rocks of different ages yields an apparent wandering of the Earth's magnetic poles. We now know that it is the crust itself, not the Earth's magnetic field, that wanders. The magnetic field wobbles to some extent, of course, and changes polarity episodically, but the field is essentially a dipole coincident with the Earth's geographic rotation axis.

Arc-Trench Gap: the geographic region between seaward oceanic deep and landward volcanic arc. It is the locus of the forearc basin. (See figures 2.7, 4.10, and 4.21.)

Archeozoic (Archean) Era: the time interval between about 3.8 and 2.5 billion years ago. (See figure 7.2.)

Asteroid: subplanetary mass of condensed matter in our solar system. Most are the size of meteorites (up to a few meters in diameter) and all are less than 1,000 kilometers across. Most are confined to orbits between Jupiter and Mars—the so-called asteroid belt. (See figure 1.1.)

Asthenosphere: the soft, plastic portion of the upper mantle on a planet directly beneath the lithosphere. On Earth, the only planet for which such a weak, ductile zone has been recognized with certainty, the asthenosphere gradually passes downward into the mantle transition zone at depths below about 220–400 kilometers. (See figure 1.7B.)

Atmosphere: the gassy envelope surrounding the solid and/or liquid outer surface of a planetary body.

Atoll: an island characterized by a concentric, fringing coral reef. (See figure 2.8.) These coral rings grow upward in the near-surface zone of the sea at about the same rate as the underlying island sinks beneath the surface of the ocean (due to lithospheric cooling and contraction).

Atom: the fundamental chemical particle that, in sufficient numbers, displays all the characteristic features of the element. An atom consists of a positively charged nucleus (protons plus neutrons) and an enveloping, negatively charged electron cloud. (See figure 3.2.)

Atomic Group: vertical column in the periodic table of the elements (see last page of book). Each neutral atom in a particular group has the same number of valence electrons but different numbers of electron orbital shells.

Atomic Nucleus: central, massive portion of an atom, consisting of protons and neutrons.

Atomic Number: known also as Z, it is the number of protons characteristic of an element, and defines its chemical nature.

Atomic Series: horizontal row in the periodic table of the elements (see last page of book). Each neutral atom in a particular series has the same number of electron shells, but a different number of valence electrons in the outermost orbital.

Atomic Weight (Mass): essentially, the sum of the masses of the protons and neutrons of an element. For instance, the common isotope of oxygen, ^{16}O, has eight neutrons, and eight protons, thus an atomic mass of 16.

Attenuation: reduction in amplitude of energy, or, in a structural context, thinning of material. Attenuation of seismic wave energy means reduced amplitude at the same frequency; attenuation of continental crust means necking down of the continent by stretching or pulling it apart.

Axis of n-Fold Symmetry: a rotation of 360 degrees/n about an n-fold axis of symmetry results in self-coincidence. For instance, a common pencil is a hexagonal prism, and possesses a sixfold axis of symmetry parallel to its length; ignoring the lettering on one side of the pencil, a rotation of 60 degrees results in an identical aspect to that prior to the rotation.

Azimuth: angular direction from north measured on or parallel to the surface of the Earth. An azimuth of N60E is a direction 60 degrees clockwise (i.e., east) of north; similarly N15W is a direction 15 degrees counterclockwise (i.e., west,) of north.

Banded Iron Formation: well-stratified sedimentary layers of moderate thickness but great lateral extent. Banded iron formations consist of chert and iron oxide, precipitated chemically in a stable shallow sea far from a source of clastic debris. Most such deposits accumulated during Early Proterozoic time. Economic iron ore deposits generally are secondary;

that is, they form subsequent to sedimentation, due to the preferential dissolution of silica by vast quantities of circulating groundwater or hydrothermal solutions.

Basalt: a dark, ferromagnesian lava (lava rich in iron plus magnesium) typical of the upper part of the oceanic crust. Basaltic flows also occur in the continental crust. (See chapter 4 and table 4.1 for description and classification.) Basalt is the common lava of the Hawaiian Islands.

Beneficiation: concentration of an ore by one or more of several mechanical and/or chemical processes.

BETA (β) Emission: the nuclear process whereby a neutron emits an electron (a negative charge) and is transformed into a proton (a positive charge).

Binding Energy: the subatomic forces that hold the atomic particles together in the nucleus.

Biochemical Sediment: sedimentary accumulation of organically precipitated material.

Bioclastic Sediment: sedimentary accumulation of fragments of shell material (produced by organisms).

Block-fault Mountains: mountains that originate, not from folding and crumpling of the constituent rocks, but by fault movements. Some blocks are elevated, others depressed as differential creep takes place at depth.

Blueschist: a lineated, low-grade, high-pressure metamorphic rock produced in subduction zones, and characterized by glaucophane, epidote or lawsonite, and hydrous ferromagnesian silicates. (See figure 4.19.)

Body Waves: vibrational energy propagated through a body—such as the Earth.

Bonding: the forces that attract atoms to one another. They are of four principal types: ionic, covalent, metallic, and van der Waals. (See figures 3.5 through 3.8.)

Borderland: a hypothetical land mass lying off the margin of a continent. Most borderlands were in actuality adjacent sialic masses that drifted against or away from the reference continent during sea-floor spreading and continental drift.

Breccia: rocks of a broad range of surface or near-surface origins, all of which are characterized by angular fragments. Without modifiers, the word commonly refers to sedimentary gravels and conglomerates that contain sharp, angular clasts. Igneous and metamorphic breccias consist of preexisting rocks that have been brittlely deformed and broken into angular pieces, due to internal flow, meteorite impact, crustal deformation, or the like. Breccias also occur along upper crustal portions of faults where frictional drag has fragmented the wall rocks.

Bridging Oxygen: in a silicate structure, an oxygen that is bonded to two tetrahedrally coordinate silicons (or one silicon and one tetrahedrally coordinated aluminum). Oxygens bound to one tetrahedrally coordinated silication but not shared between two tetrahedra are termed nonbridging oxygens (e.g., see figure 3.18B).

Calcite Compensation Depth: the depth in the ocean below which $CaCO_3$ dissolves (due to increased pressure), and therefore carbonate strata are not deposited. At present, it lies about four kilometers below sea level.

Camouflaged Element: a trace element substituting to a very minor extent for a common element in a particular structural site in an abundant rock-forming mineral, such as rubidium for potassium in orthoclase, silver for lead in galena, uranium for zirconium in zircon.

Cap Rocks: strata impermeable to fluid flow that overlie a petroleum-bearing (or other volatile-saturated) reservoir rock and thus prevent the oil and gas (or other fluids) from migrating away. (See figure 6.8.)

Carbonaceous Chondrite: a rare type of chondritic meteorite containing poorly crystallized carbon and hydrous minerals.

Cataclastic Metamorphic Rocks: metamorphic lithologies that develop from pre-existing rocks through the process of differential shear, fracturing, and mechanical milling down of the original grains without significant chemical reconstruction (recrystallization).

Cation: a positively charged ion ($Z > e$); so-called because, in an electrolyte solution, positively charged ions migrate toward the negative electrical terminal, or cathode.

Central Eruptive-type Volcano: central cone, constructed from the solidification of viscous lava and ash. Eruptions are characteristically violent. (See figures 4.2 and 4.3.)

Chalk: very fine-grained bioclastic limestone. Individual particles in many such well-bedded deposits consist of the shells of microorganisms known as foraminifera.

Chemical Sediment: accumulation of inorganically precipitated material, characteristic of quiet-water depositional environments.

Chert: sedimentary rock dominantly or exclusively consisting of biochemically or chemically precipitated, laminated silica.

Chilled Contact: the rapidly cooled or quenched margin of an igneous body; may be glassy, but in any case is finer grained than the more slowly cooled interior of the mass.

Chondrite: a stony meteorite containing abundant blebs of magnesium-rich silicates (the droplets are former molten condensates called chondrules).

Chondrule: a spherical, formerly molten droplet of silicate material in a meteorite. Generally, formerly molten or glassy chondrules have been recrystallized and consist of intergrown prismatic sprays of magnesium-rich silicates such as olivine and/or enstatite; a meteorite that contains these blebs is termed a chondrite.

Clastic (Detrital) Sediment: surficial accumulation of particulate matter or clasts (also called detritus, and detrital sediments).

Claystone: a massive, clay-rich variety of mudstone; a rock consisting of very fine (clay-size) detritus.

Cleavage, mineral: planar fractures parallel to high atomic density planes in the crystal structure of the host mineral. Micas have one perfect basal cleavage parallel to the tetrahedral sheets, whereas chain silicates possess two (prismatic) planar fractures, the intersections of which are parallel to the tetrahedral chain length.

Cleavage, rock: planar fracture surfaces in a metamorphic rock that are parallel to the shape orientations of the constituent minerals. Fracture cleavage consists of discrete planar breaks separated by intervening, undeformed rock. Flow cleavage is a three-dimensional fabric of the rock and is pervasive, penetrative. Cleavage is produced by the attendance of differential stress during recrystallization.

Coal: the accumulation of vegetal matter, chiefly trees, in a swampy environment, later compressed and heated by burial in a sedimentary basin, now consisting chiefly of carbon. Increasing temperatures and pressures during burial result in the progressive devolatilization series peat, lignite, bituminous coal, anthracite.

Coal Rank: refers to the gradual thermally induced loss of volatile constituents (N_2, H_2S, H_2, H_2O, CO_2, SO_2, etc.), from original woody materials, leaving behind more nearly pure carbon at progressively higher ranks.

Color Index: the volume percentage of ferromagnesian (dark-colored) minerals in an igneous rock. Darker rocks are called mafic, lighter rocks are termed felsic.

Columnar Joints: planar fractures that develop at right angles to the rapidly cooling surface of lavas and intrusive igneous melts injected at very shallow crustal levels. Seen on the cooling surface itself, the fractures (or joints) form polygonal—often hexagonal—prismatic sets.

Comet: a relatively small planetary body that is surrounded by a gaseous halo; its tail points away from the Sun due to the solar wind. Comets are composed chiefly of ices of the volatile elements, and have been described as "dirty snowballs." Their home is in the Oort cloud, well beyond Neptune and Pluto at the outer limits of our solar system.

Composite Igneous Body: two or more distinct magma injections of different chemical/mineralogic compositions in a single pluton.

Compound: an electrostatically neutral chemical species consisting of definite (stoichiometric) proportions of two or more elements (e.g., H_2O, $NaAlSi_3O_8$).

Compressional Mountain Belt: an orogenic belt exhibiting folded, shortened, rumpled sections of rock.

Compressional Seismic Waves: so-called P waves, transmitted by a push-pull mechanism (compression-rarefaction), rather like the propagation of energy along a row of touching balls when the first is struck with a hammer. The waves are called "P" for "primary" because these are the fastest waves that pass through the Earth and are received first by a seismograph station. (See figure 1.8A.)

Compressive Bend: constraining bend along a strike-slip and/or transform fault boundary. Commonly the site of a topographic high because of excess mass concentration. (See figure 2.19.)

Conchoidal Fracture: any curviplanar fracture, such as that characterizing the glassy lava obsidian, or the mineral quartz.

Concordant Igneous Body: a roughly tabular intrusion, the margins of which are parallel to the layering of the wall rocks. (See figure 4.5B, C, D.)

Conglomerate and Gravel: clastic sedimentary rock and unconsolidated detritus, respectively, the particles of which are coarser grained than sandstone.

Conservative Plate Boundary: a near-vertical fault or fracture zone (most common examples are oceanic fracture zones) along which lithospheric plates slide past one another but are being neither created nor consumed. They are the locus of transform faults. (See figures 2.9, 2.14, and 5.7.)

Contact Metamorphic Rocks: metamorphic rocks that develop from pre-existing rocks through the process of recrystallization localized around a hot spot, such as an igneous pluton. The zonation of recrystallization in the country rocks is called a metamorphic aureole.

Continental Drift: the differential motion of the continents, first elaborately described and detailed by Alfred Wegener. (See figure 2.2.)

Continental Rise: the broad, inclined submarine surface from the base of the continental slope (approximately three-kilometer water depths) to the deep, abyssal ocean basin (approximately five-kilometer water depths).

Continental Shelf: portion of continental crust covered by shallow seas; water depths are generally less than 200 meters.

Continental Slope: the transition zone between the base of the continental shelf (approximately 200-meter water depths) and the continental rise (approximately three-kilometer water depths).

Continental Transform Basin: a basin formed by differential motion along a transform fault transecting a portion of the continental crust, resulting in mass deficiency. Where the depression is produced at a releasing bend of a transform, it is called a pull-apart basin. (See figure 2.19.)

Convection: mass circulation of a flowing medium in the gaseous, liquid, or plastic solid state. Thermal convection is a consequence of either heating a medium at depth or cooling it at the surface too fast to achieve stready-state energy transfer through conduction, and at too low a temperature for radiative transfer to be important. As material at depth is heated, it expands, becomes buoyant, and therefore rises toward a cooling surface; cooling upper layers tend to be displaced laterally, and ultimately sink into the hotter substrate due to their greater density. Thus, a circulation pattern—convection—is established. (See figure 7.11.)

Convergent Continental (active) Margin: also called Pacific-type margin. Formed in the vicinity of a subduction zone, and on the landward side. Charactcristic of island arcs as well as continental margins. Trench sediments are offscraped and underplated to the nonsubducted plate during convergence, in the process becoming severely deformed (folded, and thrust-faulted); in contrast, sediments of the forearc basin (arc-trench gap) are virtually undeformed because they are laid down in a tectonically undisturbed setting. (See figure 4.12B, C.)

Convergent Plate Boundary: a linear feature along which a pre-existing lithospheric plate is being destroyed by sinking into the mantle beneath another (the downturn is marked by an ocean trench). Also known as a consumptive plate boundary. (See figures 2.7, 4.10, and 4.21.)

Coordination Number: in a crystal structure, the number of nearest-neighbor anions surrounding a central cation, or vice versa. (See figures 3.9 and 3.10.)

Coordination Polyhedron: the figure formed by connecting the centers of all anions surrounding a central cation in a crystal structure. Polyhedra with three, four, six, eight, and twelve sides (triangle, tetrahedron, cube or hexahedron, octahedron, and dodecahedron, respectively) are common, others much less so. (See figures 3.9 and 3.10.)

Core: the inner, central sphere of a planetary body. On Earth, the core extends from depths of about 2,900 kilometers to the center at 6,378 kilometers; its composition appears to be comparable with that of iron meteorites. (See figure 1.7A, B.)

Country Rock (Wall Rock): pre-existing rock into which molten or partly molten igneous rock is intruded.

Covalent Bonding: outer-orbital electrons are shared between neighboring atoms, the electron clouds of which thereby interpenetrate one another. Discrete, strong bonds are formed by this mutual electronic interpenetration. (See figure 3.7A, B.)

Craton (Shield): a continental nucleus, generally of Precambrian age (older than 570 million years). Such nuclei are also termed shields. Cratons constitute the ancient basement of a continent. Many consist of granite plus greenstone belts and/or high-grade metamorphic gneiss complexes, or both.

Creep: virtually continuous differential movement of contiguous portions of a solid body, across a fault or shear zone.

Crust: the outer, compositionally distinct solid rind of a planetary body. On Earth it consists of two types—oceanic crust about five kilometers thick, and continental crust approximately forty (± twenty) kilometers thick. Crust makes up the near-surface portion of the lithosphere. (See figure 1.7A, B.)

Crystal: a mineral grain bounded by planar surfaces (crystal faces) that bear a simple geometric relationship to the ordered atomic arrangement—or crystal structure. (See figure 3.1A, B.)

Crystal Defect: a discontinuity in the otherwise perfect, three-dimensional repeat periodicity of a crystal structure. Included are point defects, such as the omission of a cation or anion, line, edge, and surface irregularities where an extra row or plane of atoms is present or absent, and stacking mistakes.

Crystal Faces: planar terminations to a mineral grain that bear a simple geometric relationship to the ordered atomic arrangement. Crystal faces are typically structural planes of high atomic concentration. (See figure 3.11.)

Crystal Form: all faces of the same type and shape that share a common geometric relationship to the crystallographic axes. An example is the eight equilateral-triangle faces of an octahedron. (See figure 3.12A–G.)

Crystal Structure: the three-dimensional, systematic (ordered) atomic arrangement of a crystalline solid. Periodic, identical repeats of the local atomic arrangement occur in any direction, but in general the structure is different in different directions. Although the analogy is imperfect, crystal structure may be likened to three-dimensional wallpaper. (See figures 3.13, 3.15, 3.16, and 3.21.)

Crystalline: any solid that possesses a systematic, ordered atomic arrangement—or crystal structure.

Cyrstallographic Axis: one of three principle symmetry directions in a crystal, related to the basic atomic structure of the material. Unit cell edges define the crystallographic axes on an atomic scale

Cumulate: intrusive igneous rocks made up of layers of accumulated crystals that, due to greater density relative to the melt, sink to the bottom of the magma chamber (e.g., dunites, peridotites, pyroxenites), or being lighter, buoyantly float to the top (e.g. anorthosites).

D″ Boundary: an inferred narrow zone of marked chemical contrasts at the base of the lower mantle where temperature abruptly increases downward from that of the mantle geotherm to that characteristic of the outer core. (See figure 1.10.)

Dacite: an extrusive igneous rock compositionally intermediate between andesite and rhyolite. (See chapter 4 and table 4.1 for description and classification.)

Deep-Focus Earthquake: energy release (hypocenter) at depths of 200–700 kilometers in the Earth.

Density: mass per unit volume, usually given as g/cc (essentially the same as specific gravity).

Depleted Mantle: composition of mantle after low-melting crustal constituents have been preferentially removed through partial fusion, followed by separation and ascent of buoyant magmas.

Detritus: sedimentary particulate matter that accumulates to form detrital or clastic rocks.

Deuterium: a heavy isotope of hydrogen, containing one neutron in the nucleus (instead of none, as in normal hydrogen). An even heavier isotope of hydrogen is tritium, which has two neutrons in its nucleus. The atoms of normal hydrogen, deuterium, and tritium thus possess atomic weights 1, 2, and 3, respectively. Of course, all isotopes of hydrogen are characterized by the same atomic number (Z), namely 1.

Devolatilzation: the driving off of gaseous constituents, particularly H_2O, but the process also may include NH_3, N_2, CH_4, CO_2, CO, H_2S, SO_2, F_2, and Cl_2.

Diabase: a rock produced by very shallow-level intrusion of mafic magma. Textures are intermediate between compositionally equivalent fine-grained basalt and coarse-grained gabbro. (See chapter 4 and table 4.1 for description and classification.)

Diagenesis: the chemical and physical changes in a sediment after deposition, during burial, and before weathering. Compaction and cementation result in the formation of a solid rock through a process known as lithification. With continued burial and recrystallization, diagenesis gradually gives way to very low-grade metamorphism.

Dike: a tabular igneous intrusive, discordant (not parallel) to the layering of the wall rocks. (See figure 4.5A.)

Discordant Igneous Body: a tabular or irregularly shaped intrusion, the margins of which transect and cut the compositional layering or fabric of the wall rocks. (See figures 4.4 and 4.5A.)

Divergent Plate Boundary: a linear feature along which new lithosphere is forming and being transported away at right angles to the plate boundary (the latter is marked by an oceanic ridge). Also known as a constructive plate boundary. (See figures 2.6, 4.9, and 4.20.)

Dolomite: a fine-grained, chemically precipitated sedimentary rock formed by the settling out of carbonate on the sea floor or lake bottom. The principal phase constituting the rock dolomite is a mineral of the same name. Its chemical formula is $CaMg (CO_3)_2$. Some dolomite may represent a primary precipitate, but most such rocks evidently have been formed by the later chemical replacement of primary calcite. (See chapter 4 for description and classification.)

Double-chain Silicate: a group of minerals characterized by a pair of silicon-oxygen tetrahedral chains polymerized to infinity along a single direction and cross-bonded to the neighboring chain by every other tetrahedron. The unit repeat of the chain in $(Si_4O_{11})^{-6}$, and each tetrahedron, on the average, possesses 2.5 bridging oxygens and 1.5 nonbridging oxygens. (Of course, only half of each bridging oxygen can be assigned to a specific tetrahedron.) Some of the silicon may be replaced by aluminum. (See figure 3.19B.)

Dunite: a nearly monomineralic ultramafic rock consisting chiefly of olivine.

Dynamo: a machine or construct of nature capable of generating an electromagnetic field. The Earth's core is thought to act as a dynamo: the surrounding magnetic field is produced by virtue of convection in the outer, liquid core. (See figure 1.11.)

Dynamothermal Metamorphism: regional recrystallization typified by accompanying deformation. The metamorphic rocks characteristically consist of dimensionally aligned mineral grains.

Eclogite: a high-grade, high-pressure metamorphic rock characterized by intermediate sodium-calcium pyroxene solid solution and garnet, and totally lacking feldspar. Eclogites have been formed exclusively in the very deep continental crust, in subducted oceanic crust, and in the deep upper mantle. (See figure 4.19.)

Ejecta: particulate matter thrown out of a volcano. Depending on grain size, it is termed ash (very fine grained), lapilli (sand sized), and larger blocks (angular) or bombs (rounded).

Electron: a negatively charged atomic particle that is confined to an orbital zone—or cloud—surrounding the nucleus. It is approximately 1/1,837 the mass of a proton or neutron.

Electron Capture: the process whereby a proton and an electron combine to produce a neutron.

Electron Orbitals (Electron Shells): the disposition of electrons about a central nucleus. Referred to as an electron cloud, discrete electron shells and suborbitals represent differing energy levels and numbers of electrons statistically contained therein. (See figure 3.2.)

Electron Shells: the electron orbitals that define the energy levels and the statistical complement of electrons in the cloud surrounding the central nucleus. In the Bohr atom, these shells are K, L, M, N, which can accommodate up to two, eight, eighteen, and thirty-two electrons, respectively. (See figure 3.2.)

Electron Subshells: subdivisions of the electron shells, whose orbits are, for the most part, elliptical, not circular. Subshells are s, p, d, and f, accommodating up to two, six, ten, and fourteen electrons respectively. (See figures 3.3 and 3.4.)

Electronegativity: a quantitative measure of the ability of an atom to attract electrons. Atomic species with high electronegativities readily become anions, whereas those possessing low electronegativities becomes cations. (See figure 3.8.)

End Member: where compositional variation is possible for a mineral series, end members mark the theoretical chemical limits. For example, in alkali feldspars, $(Na,K)AlSi_3O_8$, compositional variation may be described in terms of two end members, $NaAlSi_3O_8$ and $KAlSi_3O_8$; such compositions are represented by real minerals in nature, albite and orthoclase respectively.

Epeiric Seas: the shallow oceans covering continental shelfs, and platforms, such as the Grand Banks, and the Gulf of Mexico. Water depths are generally less then 200 meters.

Epicenter: the surface location directly overlying an earthquake energy release at depth in the Earth (the latter is called the focus or hypocenter).

Equilibrium: the minimum energy configuration, in either a mechanical or chemical system.

Estimated Reserve: an educated guess at the amount of reserve for a particular natural resource (see Reserve).

Eugeoclinal Sediments: poorly sorted, poorly bedded, usually very thick, somewhat deformed sections laid down along a convergent margin in the vicinity of a trench. Chaotic, volcanogenic sediments include submarine debris flows and tectonically disrupted strata. Many such deposits accumulate in and near oceanic trenches. Faulted-in blocks and slabs of oceanic crust are common. (See figure 4.12B, C.)

Euhedral: adjective denoting a mineral grain bounded chiefly by (planar) crystal faces. (See figure 3.1A, B.)

Evaporite Deposit: laminated salt beds resulting from the evaporation to dryness of briny seawater.

Excited State of an Atom: the situation in which one or more inner-orbital electrons may reside in an outer orbital, or may be missing completely (ionic state).

Extrusive Igneous Rock: the surficial (including submarine) accumulation of magma-derived material. Extrusives include lava flows, ash falls, ash flows, etc.

Fault: any fracture in a solid geologic body across which differential motion has occurred.

Fault Breccia: materials composing the walls of a fault zone that have been broken into angular fragments during differential movement (confined to the upper crust where rocks behave brittlely).

Fault Gouge: unconsolidated, fine-grained, milled down material along a shallow-level fault zone. The material is commonly claylike in its physical behavior.

Felsic Igneous Rock: magmatic rock unit rich in felsic constituents, silicon, sodium, and potassium, and generally poor in ferromagnesian constituents, iron, magnesium, and calcium.

Fertile Mantle: mantle that still contains crust-forming elements. Also often called primitive or undepleted mantle.

First-cycle Sediment: a clastic sediment, the grains of which were mostly derived from a pre-existing igneous or metamorphic rock, and which exhibits only limited effects of sedimentary differentiation (the latter results from prolonged weathering and transportation).

Fissure-Type Volcano: extrusive deposits erupted quiescently along a fracture; such volcanoes have low, broad profiles and are chiefly constructed of highly fluid, low-silica lavas.

Foliation: the preferred orientation of platy mineral grains in a metamorphic rock that imparts to it a layered structure and a ready cleavage (foliation). (See figure 4.13A, B.)

Forearc Basin (Marginal Basin): a structural trough situated outboard (seaward) from a volcanic continental margin or island arc, and inboard (landward) from an oceanic trench complex. The site of the forearc basin is sometimes referred to as the arc-trench gap portion of a convergent plate junction. (See figures 2.7, 4.10, and 4.21.)

Foundering of Crust: the sinking of a solid, relatively high-density crust through an underlying, less dense, molten, or partly molten, substrate.

Fracture, Mineral: any breakage surface that does not coincide with a plane of atoms in the rigorous, three-dimensional atomic arrangement (or crystal structure) of the mineral.

Fracture, Rock: any planar surface of breakage in a rock; may belong to a set of parallel cleavage surfaces.

Framework Silicate: a group of minerals characterized by cross-linkage of silicon-oxygen tetrahedra to infinity in three directions. The unit repeat of the framework is SiO_2, and each tetrahedron has four bridging oxygens. (Of course, only half of each bridging oxygen can be assigned to a specific tetrahedrdon.) For framework minerals other then the SiO_2 polymorphs, a portion of the silicon is replaced by aluminum, thus requiring the presence of additional cations for charge balance. (See figure 3.21.)

Freeboard: that portion of a material rising above sea level. As continents attain increasing emergence, the agents of erosion tend to intensify, degrading the crust towards sea level.

Fusion Power: the combination of ionized hydrogen atoms (including heavier isotopes such as deuterium) at extremely high temperatures. Fusion produces helium nuclei, liberating neutrons and vast amounts of energy in the process. This is the process that fuels the sun.

Gabbro: a dark, ferromagnesian rock, cooled at depth from magma of basaltic composition, typical of the lower part of the oceanic crust. Gabbroic intrusions also occur in continental crust. (See chapter 4 and table 4.1 for description and classification.)

Geochronology: quantitative study of the age of earth materials utilizing radioactive decay of elements in mineralogic and lithologic systems as "atomic clocks"; the transmutations are characterized by measured, constant rates of atomic transformation independent of temperature, pressure, or compositional concentration. Measurement of the concentrations of parent and daughter isotopes thus yields information concerning the age of formation of the system.

Geomagnetic Time Scale: a measure of the geologic past history of normal and reversed magnetic fields as a function of time; this history is contained in the remanent (remnant) magnetism of rocks. (See figures 2.11 and 7.2.)

Geothermal Energy: an energy resource resulting from the flow of heat within the Earth toward the cooling surface of the lithosphere. Anomalously high heat-flow regions are utilized for the energy production.

Geothermal Gradient (Geotherm): the increase of temperature within the Earth as a function of depth, usually given in units of degrees Celsius per kilometer. (See figures 1.10 and 7.3.)

Geyser: episodic or periodic conversion into steam of surface and/or groundwater draining into a constricted fissure in hot rock. The wall rock, of course, has to be hotter than the boiling point of H_2O for geysering to occur.

Glossopteris Flora: a distinctive, latest Paleozoic group of fossil plants geographically confined to the Gondwana (southern hemispheric) supercontinent. (See figure 2.3.)

Gneiss: a fairly coarse-grained, foliated metamorphic rock characterized by heterogeneous layering. The foliation is called gneissosity. (See figure 4.13C.)

Gondwana: the Late Paleozoic assembly of South America, Africa, India, Australia, and Antarctica, at times separated from the other great continental mass (Laurasia) by a narrow seaway (Tethys). (See figures 2.1 through 2.4)

Granite: a highly silicic, alkali-rich rock, cooled at depth from magma approximating rhyolitic in composition, and typical of deeper parts of the continental crust. (See chapter 4 and table 4.1 for description and classification.) Often the terms "granite" and "granitic" are applied more loosely to rocks of the quartz diorite, granodiorite, and quartz-monzonite groups, as well as to granite in the narrow sense.

Granitization: the metamorphic conversion of pre-existing rock to a more granitic texture, mineralogy, and bulk composition: accomplished by chemical exchange with a hot aqueous fluid in the absence of partial melting.

Granite plus Greenstone Belt: a linear downwarped zone of upper crustal, low-grade metamorphosed basalts and interlayered volcanogenic sediments, intruded around the margins by more buoyant granitic plutons. Characteristic of the Archean rock record. (See figures 7.8 and 7.9.)

Greenhouse Effect: the heating of the atmosphere and hydrosphere due to the gradual buildup of (anthropogenic) CO_2 and related gas species (such as Ch_4 and N_2O). These gas molecules absorb incoming solar radiation, which in turn is re-emitted at different frequencies and trapped by the atmosphere, resulting in global warming.

Greenhouse Gas: one of several molecular species, the atomic modes of vibration of which are appropriate for the absorbance of sunlight. Examples include CO_2, N_2O, methane, and man-made chloroflourocarbons, carbon dioxide being the most concentrated and that which most rapidly builds up in the Earth's atmosphere.

Greenschist: a foliated low-grade metamorphic rock characterized by albite, chlorite, epidote, and actinolite. (See figure 4.19.)

Greenstone: a term describing a green-colored, low-grade metamorphic rock, the protoliths of which are chiefly (but not exclusively) ferromagnesian igneous rocks such as basalt. (See chapter 4 for descriptions and classification.) Most greenstones are massive, not foliated or lineated; many contain the same minerals as greenschists.

Greenstone Belt: a long, narrow lithotectonic zone consisting of feebly recrystallized basalts and interlayered volcanogenic sedimentary rocks. The rocks are green because of the abundant presence of chlorite, a hydrous, ferrous aluminosilicate. Other associated and abundant green minerals are actinolite and epidote. (See figures 7.8 and 7.9.)

Ground State of an Atom: the lowest energy configuration (electron population) of the electron orbitals.

Groundwater: water residing beneath the surface of the solid earth in fractures and in the interstices of porous rocks and soil. Its upper surface (with unsaturated, porous rock and soil) is known as the water table. Wells drilled for the purpose of extracting groundwater must penetrate the water table in order to reach and withdraw water.

Guyot: a flat-topped submarine mountain. This bathymetric feature is produced by igneous eruption that results in construction of a conical edifice rising above sea level during its active stage; subsequently the mountain is planed off by postvolcanic erosion approximately to sea level. Next, regional subsidence of the entire structure and subjacent sea bottom (thought to be due to gradual cooling and thickening of the lithosphere) causes the beveled, truncated cone to sink beneath the seas. (See figure 2.8.)

Hadean Time: the interval between the formation of the Earth and the beginning of the Archean, from about 4.5 to 3.8 billion years ago. (See figure 7.2.)

Half-life: the time required for half the amount of reactant present to be converted to product assemblage. Most commonly employed to describe the rate of decay of a radioactive element to its daughter products. (See figure 7.1.)

Hardness: a substance's resistance to abrasion. Harder minerals readily scratch softer minerals, but not vice versa. Mohs' scale of hardness for minerals is presented as table 3.2.

Heat Conduction: transmission of thermal energy by propagation of atomic vibrational energy through a medium: most effective in condensed (solid and liquid) systems.

Hemipelagic Sediment: mixed, very fine clastic and chemically/biochemically precipitated debris drifting downward and deposited as a thin blanket on the deep ocean floor: the so-called layer 1 of oceanographers. Consists principally of finely laminated silica and/or carbonate beds ± manganese nodules. (See figures 2.6 and 4.9.)

High-Grade Metamorphic Gneiss Complex: a broad region of deep- or mid-crustal, intensely recrystallized, interlayered mafic and felsic rocks that may represent the deep roots of mountain ranges, island arcs, and/or granite plus greenstone belts. Characteristic of the Archean rock assemblage. (See figures 7.7, 7.8, and 7.9.)

High-Grade Ore: a rich, highly concentrated ore deposit. Even small-volume occurrences can be worked profitably.

Hornfels: a type of contact metamorphic rock characterized by a lack of deformation, or strain; such rocks do not exhibit foliation or cleavage, and instead possess a homogeneous, mosaic texture of interlocking, equidimensional grains.

Hot Spot: the deep upper mantle source of a rising column, or plume, of hot —and therefore buoyant—mantle material. The thermal anomaly seems to be stationary in contrast to the overlying, moving lithosphere. (See figure 5.4A, B, C.) Could hot spots be produced by impacting asteroids? It is possible, because of the local pressure release resulting from cratering, but not known for certain.

Hydrocarbons: any of a family of organically produced hydrogen-carbon compounds. Coal is a mixture of solid, oil of liquid, and natural gas of gaseous hydrocarbons.

Hydrosphere: the liquid sheath, or layer, situated between solid and gaseous portions of a planetary body. On Earth, it consists of the world oceans, rivers, and lakes.

Hydrostatic Head: the height to which an unconstrained column of fluid (e.g., water or magma) will rise in the Earth.

Hydrothermal Fluid (Solution): hot aqueous fluid derived at depth from a local heat source (typically a congealing magma), which moves into the surrounding rocks, reacting with and leaching them in the process. As the H_2O-rich fluid cools, dissolved substances (such as quartz, carbonates, and metal-bearing sulfides) precipitate out, forming veins and disseminated mineralization. The associated process of chemical change in the host rocks is known as metasomatism.

Hypocenter: the actual location—or focus—of an earthquake at depth within the Earth.

Igneous Fractionation (Differentiation): the partitioning of elements between melt and crystals, followed by separation of the liquid from the solids. This process results in changing composition of late-stage melts. Other differentiation mechanisms for igneous systems include liquid immiscibility, thermally driven diffusion, and assimilation of wall rocks.

Igneous Rock: any rock formed from the cooling of a molten (or partly molten) solution of silicates ± carbonates, i.e., magma. Two broad categories of igneous rocks are recognized, depending on where in the crust they have solidified: (1) surficial (lavas, ash, etc.), or extrusive; and (2), deep (batholiths, dikes), or intrusive.

Impact Breccia: pulverized angular rock fragments produced by the collision of a planetary body with a meteorite, asteroid, or comet.

Incipient Melting: the very low degree of partial fusion of a pre-existing solid.

Inclination (Dip): the maximum angle measured in a vertical plane between the horizontal and a magnetic field line (the north-seeking end of a compass), or between the horizontal and any other linear or planear feature in a rock.

Inclusion: a small piece of country (wall) rock completely engulfed by magma. (See figure 4.4.)

Index Mineral: a phase that indicates the intensity—or grade—of metamorphism. Typical index minerals include laumontite, chlorite, biotite, garnet, staurolite, cordierite, jadeite, kyanite, sillimanite, and andalusite. (See figure 4.18.)

Individual Tetrahedron: $(SiO_4)^{-4}$ tetrahedral groups that are not linked to other silicon-oxygen tetrahedra, but instead share their oxygens with other, higher coordination cations. (See figure 3.18A.)

Infall: used in reference to the melting of dispersed iron metal in the primitive, newly accreted Earth, and migration downward of the dense, coalescing metallic droplets to produce the iron (plus nickel) core.

Inner Core: the inner, spherical, central portion of the Earth's iron-nickel core, thought to be solid. (See figure 1.7A, B.)

Intermediate-Focus Earthquake: energy release (hypocenter) at depths of 50–200 kilometers in the Earth.

Interstitial Solid Solution: a mineralogic solid solution series in which chemical variation is accomplished by the addition of another atom or ion to a previously unoccupied structural site. An example is steel: here, the addition of small amounts of carbon interstitial to the much larger-scale packing of iron atoms converts the somewhat softer, more brittle iron into harder, more flexible steel.

Intrusive Igneous Rock: magma-derived material that congeals at depth within the Earth. Intrusives include stocks, batholiths, and other irregularly shaped plutons, as well as tabular sills and dikes. (See figure 4.4.)

Ion: an atom charged by virtue of its loss or gain of electrons. In the neutral atom, the number of protons, Z, equals the number of electrons, e, but for ions, $Z \neq e$. (See figure 3.6.)

Ionic Bonding: attractive atomic forces resulting from unlike neighboring ionic charges. Electrons are donated or acquired in order for the ions to achieve the noble gas electronic configuration (filled electron orbitals). General cohesiveness is a result of the mutual attraction of unlike changes, but discrete bonds are not present. (See figure 3.6.)

Iron Meteorite: any meteorite composed chiefly of an iron ($\pm$ nickel) metal alloy.

Isograd: a line on a map (or a planar surface in the Earth's crust) that locates the first appearance with advancing metamorphic grade of a particular index mineral. (See figure 4.14.)

Isostasy: the gravitative adjustment of volumes of the Earth's crust and uppermost mantle whereby all vertical columns down to a depth of compensation—say, the top of the asthenosphere—have the same mass-per-unit cross section (surface area). (See figure 1.12.)

Isotopes: where the number of neutrons associated with a fixed number of protons (Z) for an element is variable, the different atomic species are characterized by different masses, and are referred to as isotopes. For instance, the heavy isotope of oxygen, ^{18}O, has ten neutrons and eight protons, thus an atomic mass of 18, whereas the common isotope, ^{16}O, possesses eight neutrons and eight protons.

Kimberlite: a rare variety of igneous rock derived from the mantle, and consisting chiefly of hydrous minerals such as serpentine, but also typically containing calcium and/or magnesium carbonates. Kimberlites rise from the mantle in a fluidized condition by virtue of the presence of abundant associated volatile constituents, especially H_2O and CO_2. On ascent and decompression, the expanding gases propel the fluidized material surfaceward at an ever-increasing rate. Kimberlites are famous because some contain trace amounts of diamond. And, because diamonds are produced in the Earth only at pressures exceeding about forty kilobars, this tells us that kimberlites have risen from mantle depths of at least 125 kilometers.

Kimberlite Pipe: an upward-enlarging conduit filled with kimberlite, a fluidized igneous rock derived from the upper mantle; such H_2O- and CO_2-rich complexes ascend into the continental crust in a relatively cool state due to the expansion of the associated volatiles.

Komatiite: ultramafic lava carrying major amounts of olivine; the bulk composition of the lava is similar to that of undepleted mantle peridotite. Komatiites are found almost exclusively in Archean crust. (See chapter 4 and table 4.1 for description and classification, chapter 7 for thermotectonic significance.)

Land Bridge: continuous land or a string of islands that span an ocean basin and link otherwise mutually isolated continents. The Isthmus of Panama is an example of a real land bridge, but many hypothetical land bridges evoked by paleontologists to explain faunal and floral similarities in now geographically dispersed regions are no longer required because of the discovery and acceptance of continental drift.

Laterite: the conversion of pre-existing aluminous or ferruginous rocks to bauxite (aluminum ore) or hematite (iron ore) through the preferential solution leaching away of all other constituents by tropical weathering and groundwater circulation.

Laurasia: the Late Paleozoic assembly of North America and Eurasia; at times separated from the other great continental land mass (Gondwana) by a narrow seaway (Tethys).

Limestone: a fine-grained, chemically precipitated, layered sedimentary rock formed by the settling out of calcite on the sea floor or lake bottom. (See chapter 4 for description and classification.)

Lineation: a preferred orientation of an assemblage of prismatic (pencil-like) mineral grains in a metamorphic rock; this alignment imparts a linear fabric to the rock. (See figure 4.13B.)

Lithification: that process of compaction, dewatering, and cementation that coverts an unconsolidated sediment into a sedimentary rock. Also known as diagenesis.

Lithosphere: the outer rind of a planet that behaves as a coherent, relatively rigid, brittle unit. The Earth's lithosphere consists of crust plus upper-most mantle. It extends to depths of 50–200+ kilometers and is divided into several enormous segments, or plates, that move differentially relative to one another. (See figures 1.7A, B, and 2.13.)

Lithospheric Plate: a relatively rigid, coherent slab of the Earth's crust and uppermost mantle (50–200+ kilometers thick, and hundreds or thousands of kilometers in lateral, or surface dimensions), which moves differentially with respect to other plates. (See figures 1.7A, B, and 2.13.)

Lithostatic Pressure: the pressure-per-unit cross section generated by an overlying column of rock. In low-density sedimentary crustal sections, the lithostatic pressure increases downward by about a kilobar per four kilometers of overburden, whereas in the higher density upper mantle, a kilobar increment in lithostatic pressure results from slightly in excess of a three-kilometer increase in depth.

Low-Angle Reverse Fault (Thrust Fault): a low-angle (less than 15–20 degrees) planar break in which the overlying, hanging-wall has moved upward along the fault relative to the underlying footwall, generally resulting in juxtaposition of older rocks above younger rocks. These are especially abundant in accretionary prisms and subduction complexes. (See figure 2.7, 4.10, and 4.21.) Horizontal shortening is a consequence of this type of faulting.

Low-Grade Ore: a mineral deposit lean enough to be near the break-even point for profitability. To be worked successfully, such deposits generally must be very large before economies of scale are achieved.

Lower Mantle: the poorly understood, deep portion of the Earth's mantle, located at depths between approximatley 670 and 2,900 kilometers (See figure 1.7A, B.) It may be richer in iron/magnesium ratio than the overlying (and therefore less dense) upper mantle.

Lunar Highlands: elevated light-colored areas of the heavily pockmarked (cratered) surface of the Moon, characterized by anorthosite breccias. (See figure 1.4B.) This is the oldest part of the Moon, with the ages of some lunar highland rocks approaching 4.4-4.5 billion years old.

Lunar Maria: topographically subdued, basinal, dark-colored areas of the lightly pockmarked (cratered) surface of the Moon, characterized by basaltic lava flows predominantly younger than the rocks of the lunar highlands. Ages of the mafic flows are in the range of 3.0-3.9 billion years old.

Mafic Igneous Rock: magmatic rock rich in ferromagnesian constituents iron, magnesium, and calcium, and generally poor in silica and alkalies.

Magma: a molten (or partly molten) solution of silicates ± carbonates that, when congealed, forms a rock (see Igneous Rock). Magma extruded onto the Earth's surface is called lava.

Magma Ocean: refers to the theorized existance of a near-surface molten silicate layer on the primitive Earth or Moon, or both. Such magma oceans would have solidified early in Hadean time. (See figure 7.2.)

Magnetic Anomaly: magnetic field strength and orientation that differ from the regional field, either in intensity or attitude, or both. A local magnetic intensity exceeding the present regional field is termed a positive magnetic anomaly. (See figure 2.10.)

Magnetic Field Lines: the direction of a magnetic force field. On the surface of the Earth, field lines coincide with the direction a magnetic compass needle points. (See figure 1.11.)

Magnetic Reversal: the phenomenon whereby the polarity of the Earth's magnetic field reverses orientation. Field lines change orientation by about 180 degrees during a switch in polarity. (See figures 2.11 and 7.2.)

Mantle: the silicate shell of a planetary body directly beneath the crust, and extending to the outer boundary of the metallic core. On Earth, the mantle extends from beneath the Moho (the base of the crust) to depths of approximately 2,900 kilometers (the top of the core); its chemical composition appears to be comparable to that of stony meteorites. (See figure 1.7A, B.)

Marble: a coarse-grained carbonate-rich metamorphic rock, the precursor of which was generally limestone or dolomite. (See chapter 4 for description and classification.)

Mass Extinction: the sudden, complete, and ubiquitous mortality of many species of plants and/or animals due to an overwhelming environmental crisis. The rapid, cataclysmic change in environmental conditions may be brought about on Earth, for instance, through collision with a large meteorite or comet, or by widespread, pyroxysimal volcanic eruptions.

Maturity of Sediments: repeated cycles of sedimentary differentiation are required to produce diverse, chemically concentrated deposits, so first-cycle sediments are chemically almost unfractioned (immature), whereas multicycle sediments are strongly differentiated compositionally (mature).

Melange: a chaotically deformed sedimentary rock. Two types are common: (1) submarine slump deposits, or olistostromes; and (2) strata folded, sheared, and disaggregated in a subduction zone by deformation, or tectonic melanges. The former occur on the sea floor at low confining pressures, whereas the latter are produced under conditions of considerable overburden.

Metallic Bonding: weak cohesive force characteristic of metals, whereby positively charged metal ions are bonded together by a loosely configured, negatively charged swarm of electrons, or electron "gas," which permeates the material. High mobility of the latter is responsible for the great electrical and thermal conductivities of metals.

Metamorphic Aureole: zonal development of newly generated metamorphic minerals in country rocks around a hot spot (generally a pluton).

Metamorphic Grade: the relative intensity of recrystallization, correlated with physical conditions: the higher the grade, the higher the inferred pressure and, especially, the temperature that attended the metamorphism. (See figures 4.18 and 4.19.)

Metamorphic Mineral Facies: an assemblage of minerals, commonly defined in rocks of basaltic composition, that reflects the relative intensity of metamorphic recrystallization. (See figure 4.19.)

Metamorphic Rock: any rock the pre-existing phases and/or textures of which have been modified by solid-state recrystallization and/or deformation with or without the interaction of a fluid phase—usually taking place at depths in the Earth. Three chief types are recognized: (1) contact metamorphic rocks, disposed about a thermal anomaly such as an igneous intrusion; (2) cataclastic rocks, milled down but not recrystallized during near-surface deformation; and (3) regional, or dynamothermal metamorphic rocks, generally deformed and recrystallized over a broad region, and in many cases unrelated to igneous activity.

Metasomatism: a metamorphic process whereby the bulk chemical composition of an original rock is altered, generally by exchange with an infiltrating hydrothermal solution charged with alkalies and silica.

Migmatite: a deeply buried rock of the continental crust characterized by recrystallized solid but plastic, refractory ferromagnesian layers, and felsic layers that crystallized from a melt, or melt plus crystals. Many migmatites form by ultrametamorphism from the in-place partial fusion of a pre-existing rock.

Mineral: a naturally occurring, inorganically grown solid that has a fixed chemistry, or limited range in composition between end members, and a rigorous, three-dimensional atomic order, or structure.

Mineralogic Barrier: where the concentration of a sought-after element is below the saturation limit (the mineralogic barrier) of common rock-forming minerals, the element is broadly dispersed throughout these phases (i.e., it is camouflaged), and is not easily concentrated. When the abundance in a rock of the element in question increases, the constituent minerals become saturated and at yet higher concentrations, a separate phase containing large amounts of the sought-after element forms. The latter, typically an oxide, sulfide, or, in some cases, native element, may be of economic significance. (See figure 6.3.)

Mineraloid: any naturally occurring solid or liquid that lacks a long-range ordered atomic arrangement—or crystal structure.

Miogeoclinal Sediments: well-bedded, usually thick sections laid down along a passive (Atlantic-type) margin or in the forearc of an active (Andean-type) margin on the continental shelf, slope, and rise. Igneous rocks are not commonly associated with such sedimentary units. (See figure 4.12A.)

Mohorovicic Discontinuity: also known as the Moho or M discontinuity, it is the horizontal or subhorizontal boundary between crust and mantle. It represents an abrupt change in physical properties such as density, seismic energy transmission velocity, and composition of the materials; it may be vertically gradational over a kilometer or two beneath the continents, but is generally much sharper than that, especially at the base of the oceanic crust. (See figures 1.7A, 2.6, 2.7, etc.)

Molecule: complete, stable combinations of two or more atoms bonded together; e.g., H_2, O_2, CH_4, S_8. (See figure 3.7.) The above are chiefly gas molecules. Molecules, as such, do not exist in minerals because of the three-dimensional bonding in minerals. You cannot, for instance, recognize an NaCl molecule in the structure of halite, nor an SiO_2 molecule in quartz.

Moment of Inertia: defined as the product of the mass (M) times the square of an effective radius (r), times a constant that is related to the distribution of matter in a rotating body. For a homogeneous sphere, the moment of inertia is $0.4 Mr^2$. Planets with their masses strongly concentrated internally (in massive, metallic cores) such as the Earth and Jupiter are characterized by moments of inertia of $0.33 Mr^2$ and $0.26 Mr^2$, respectively.

Mudstone: a fine-grained, nonlaminated, massive clay-rich clastic sedimentary rock formed by the accumulation of tiny mineral flakes and grains, generally in a marine basin or lake. (See chapter 4 for description and classification.)

Multicycle Sediment: a clastic sediment, the grains of which have repeatedly undergone cycles of weathering, transportation, abrasion, and deposition. Such rocks display marked effects of sedimentary differentiation, and are classified as mature.

Mylonite: a very fine-grained metamorphic rock, milled down by shearing (the process is called cataclasis) but not recrystallized (chemically reorganized). Mylonites are characteristic of fault and shear zones in the upper, brittle crust.

Natural Resource: naturally occurring material, including energy, useful for supporting life. A resource is considered to be the total inventory of the material or energy source in question, whether recoverable or not.

Negative Magnetic Anomaly: a local magnetic field less than the regional field strength. (See figure 2.10.)

Neutron: an atomic nuclear particle possessing approximately unit mass and no net electrical charge. Differing numbers of neutrons for an element (fixed number of protons) yield differing masses of the several isotopes of the element.

Nonbridging Oxygen: in a silicate structure, an oxygen ion bonded to one fourfold coordinated silicon (or aluminum) and one or more nontetrahedrally coordinated cations. (See figures 3.18, 3.19, and 3.20.)

Nonrenewable Resource: a resource of finite abundance—i.e., one which is not being continuously generated—such as gold, uranium, iron, and SiO_2, or is being generated at a rate infinitesimal compared with consumption, such as the fossil fuels, oil, gas, and coal.

Normal Fault: a high-vertical angle fault in which the upper block—or hanging wall—has moved down relative to the lower, footwall block. (See figure 2.18A.) Horizontal extension is a consequence of this type of faulting.

Normal Magnetic Polarity: the situation in which the Earth's determined magnetic field is similar to that of the present, whereby the north-seeking end of a compass needle points north. (See figure 2.11.)

Nuclear Fission: the splitting of heavy (high atomic number) elements such as uranium, thorium, and plutonium into lighter, lower-Z (lower atomic number) daughter products. A large amount of energy is released in the process. This type of reaction was utilized in atomic bombs, and is employed today in conventional nuclear power plants.

Nuclear Fusion: the combination, or "burning" of light elements (those with low atomic number) such as hydrogen, deuterium, and helium, to produce heavier, higher-Z (higher atomic number) elements. An extremely large amount of energy is released in the process. Nuclear fusion is characteristic of stars and our Sun as well as the hydrogen bomb.

Nucleus: the massive central core of an atom, consisting of protons and neutrons.

Nuée Ardente: a glowing (incandescent) avalanche cloud of volcanic ash (see ejecta).

Oceanic Deep (Trench): an especially deep linear trough in an ocean basin (water depths characteristically are six to eight kilometers or more). These negative bathymetric features mark the locus of downturn of lithospheric plates as the latter descend into the deeper mantle at convergent plate junctions. (See figures 1.6, and 5.1.)

Oceanic Fracture Zone: vertical planar features at right angles to oceanic ridges; the fractures are marked by submarine escarpments. These zones (transform faults) are caused by differential motion of lithospheric plates. (See figures 2.5, 2.9, and 2.14.)

Oceanic Rise (Oceanic Ridge): a submarine mountain chain of great length; it is the bathymetric reflection of upwelling convective currents in the asthenosphere. Near-surface cooling produces a pair of capping lithospheric plates that are continuously transported away from the ridge axis or linear spreading center. (See figures 1.6, 2.5, and 5.1.)

Oceanic Trench (Oceanic Deep): a linear submarine depression well below the general level of the sea-floor—the bathymetric expression of a down-turning oceanic crust–capped lithospheric plate that is descending into the deeper mantle in a subduction zone. (See figures 1.6 and 5.1.)

Oil Pool: a concentration of petroleum within the pore space (interstices) of a reservoir rock, typically a porous sandstone or a fractured or cavernous limestone. Being of lower densities, natural gas occupies the highest portion of the stratal reservoir, the liquid petroleum being beneath, but overlying the yet denser water. (See figure 6.8.) In general the petroleum has migrated from a source rock—typically a shale.

Oil Shale: a muddy, stratified sediment in which the original organic matter has neither thermally matured (has not been distilled) nor migrated away.

Omission Solid Solution: a mineralogic solid solution series in which chemical variation is achieved by the systematic absence of a small proportion of the constituent atoms—either cations or anions. An example is pyrrhotite that has the nominal formula FeS; a small cation deficiency in Fe in natural pyrrhotites is accommodated by oxidation of minor Fe^{+2} to Fe^{+3}, or possibly by a comparable reduction of S^{-2}.

Oort Cloud: the feebly condensed region at the outer limit of our solar system, populated by swarms of comets, tenuous gas clouds, and dispersed ices.

Ophiolite: a slice of oceanic crust and its upper-mantle underpinnings, stranded against sialic crust by tectonism. An undisturbed, complete ophiolite section consists of overlying deep-sea sediments (chert and/or deep-water limestone) surmounting basaltic pillow lavas, breccias, and massive flows; these in turn overlie gabbros cut by diabase dikes and pass downward to chemically depleted mantle peridotites. (See figures 2.6, 4.9, and 4.20.)

Ordering: in mineralogy, a term referring to the confinement of atoms to specific sites in a crystal structure. As temperature increases, a less systematic, less ordered, more nearly random distribution of the atoms in the various structural sites (disorder) takes place.

Ore (Mineral) Deposit: a natural resource sufficiently concentrated above the terrestrial average composition so that it can be profitably extracted.

Orogenic Belt: a mountain belt in structure, which may be, but is not necessarily, reflected in the topography.

Orogeny: the process of producing structural (not necessarily topographic) mountains. This complex of structures is thought to result from compressive and tangential forces accompanying differential plate-tectonic motions, especially along and near convergent and on-land transform lithospheric plate junctions. (See figure 2.17, for example.)

Outer Core: the outer shell of the Earth's iron-nickel core, thought to be molten. Circulation (convection) within this metallic liquid alloy may be responsible for the Earth's magnetic field. (See figures 1.7A and 1.11.)

Outgassing: loss of volatiles from profound depths within the Earth because of high temperatures.

Overturned (Nappe or Recumbent) Fold: a fold in which both limbs are inclined (i.e., dip) in the same direction.

Paried Metamorphic Belts: two contemporaneously produced metamorphic zones, an oceanward high-pressure recrystallized blueschist belt, and a continentward high-temperature recrystallized, andesitic/granitic igneous belt. The narrow, seaward, linear metamorphic terrane is thought to have formed in a subduction zone, whereas the broad landward belt is regarded as having been produced in a magmatic arc at the continental margin, or in an off-shore volcanic arc.

Paired Tetrahedra: the linking of two $(SiO_4)^{-4}$ tetrahedra to produce a $(Si_2O_7)^{-6}$ complex. One oxygen—a so-called bridging oxygen—is shared by the two tetrahedra. (See figure 3.18B.)

Paleolatitude: the angular distance in degrees from the equator for a specific portion of the Earth's crust at the time of consideration in the geologic past. Paleolatitude generally is determined by studying rock remnant magnetism and/or climatic conditions recorded by contained fossils.

Pangea: the Late Paleozoic assembly of all major continental masses (Gondwana plus Laurasia) into a single supercontinent. (See figure 2.2.)

Partial Fusion: partial liquification of a pre-existing rock, whereby the lowest-melting, fusible constituents are concentrated in the melt. (See figure 7.3.)

Pegmatite: a very coarse-grained felsic, silicic, intrusive igneous rock comparable in chemistry and mineralogy to granite. Pegmatities commonly occur as irregular or tabular bodies in granite as the latest stage, lowest temperature, fluid-charged (boron-, hydroxyl-, flourine-, and chlorine-rich) crystallizing melt.

Peridotite: a very magnesian, low-silica rock rich in ferromagnesian constituents, characteristic of the upper mantle. Peridotite is similar in chemistry to stony meteorites. It is an ultramafic rock containing essential olivine plus pyroxenes. (See chapter 4 and table 4.1 for description and classification.)

Petroleum: a mixture of liquid and/or gaseous hydrocarbons produced dominantly by the distillation of the remains of microscopic aquatic plant and animal life—typically in a marine environment, much less commonly in freshwater lakes.

Phanerozoic Era: geologic interval from the end of Precambrian time, 570 million years ago, to the present (Phanerozoic time). (See figure 7.2.)

Phase: a specific state of aggregation, such as gas, liquid, or solid. Solid phases include minerals, glasses, and related substances. Rocks consist of one or more solid phases.

Physiographic Features: the topography or surface aspect of the dry land. Underwater analogues are referred to an bathymetric or hydrographic features.

Pillow Lava: ellipsoidal, bulbous blobs of basaltic lava, about a meter in average dimension, extruded and chilled under water—reminiscent (to the crazed field geologist) of pillows.

Placer: a deposit of high-density clastic grains in a stream or beach gravel; placers are characterized by settling out of particles under conditions of high flow rate of the carrier medium, water. Many gold deposits are in placers.

Plane of Symmetry: a reflection across a mirror of plane that results in self-coincidence. For instance, an automobile and a dog each possess a vertical, longitudinal plane of symmetry—at least as far as their external morphologies are concerned, but let's not get into this one again.

Planetesimal: any of a group of relatively small planetary bodies, including asteroids and meteorites.

Plasma: an incandescent ultra-high-temperature ionized gas, such as occurs in the interior of the Sun.

Platelet Stage: the Archeozoic Era in Earth history, typified by the rapid circulation of numerous, small mantle convection cells. Crust produced consisted of small, mafic island arcs, swept together in circumoceanic areas to considerable thickness by the spreading process, and later invaded by granitic melts derived by partial fusion of their lowest portions during subduction. (See figure 7.11B.)

Platform (Epeiric) Sediments: thin-bedded, areally continuous, well-sorted, and chemically differentiated (mature) sandstones, shales, and carbonate strata laid down on the stable craton or shallow-water continental shelves (epeiric seas).

Plug: a solid body of igneous rock that has cooled and congealed in a cylindrical volcanic conduit. Many plugs are thought to represent the throats of volcanoes. (See figure 4.6.)

Plume: in plate-tectonic theory, plumes are rising columns of hot mantle material, derived from a deep mantle source, or hot spot. (See figure 5.4A, B, C.)

Pluton: an intrusive igneous body of irregular or unknown shape. (See figure 4.4.)

Polarization of an Ion: distortion of the cloud of electrons surrounding an atomic nucleus from a highly symmetric, spherical configuration to a more oblate or prolate (pumpkin-like or football-like, respectively) shape, due to the attractive forces of a neighboring atomic nucleus.

Polymerization: the linking together of atomic, molecular, or ionic substructures (or polyhedra) to form a larger structural array. Pyroxenes contain infinite silicon-oxygen tetrahedral chains, and are said to be polymerized in one direction (see figure 3.19A); in contrast, silicon-oxygen tetrahedra in quartz are cross-linked in three directions (see figure 3.21); thus the structure is described as a three-dimensional framework (polymerization is in three directions).

Polymorphs: minerals having the same chemical composition but different crystal structures. Well-known examples include the polymorphs of carbon, diamond, and graphite (see figure 3.13 and 3.14), and the polymorphs of silicon dioxide, α quartz, β quartz, tridymite, cristobalite, coesite, and stishovite. Because of their contrasting three-dimensional atomic arrangements, polymorphs have different physical properties such as hardness, density, melting point, refractive index, etc.

Porphyry Copper Deposit: shallow-level portions of granitoid rocks characterized by the presence of widely disseminated, low-grade copper sulfide minerals. Such occurrences are especially typical of the western cordillera of North and South America, and of Indonesia. (See figure 6.9.)

Positive Magnetic Anomaly: a local magnetic field that exceeds the regional field strength. (See figure 2.10.)

Precambrian Shield: the ancient nucleus of a continent—a Precambrian craton. (See figure 7.10.)

Precambrian Time: Hadean, Archean, and Proterozic Eras, or the time interval from the formation of the Earth, about 4.5 billion years ago, to 570 million years ago. (See figure 7.2.) On Earth, no rocks formed during Hadean times have yet been recognized.

Precursor Phenomena: physical changes that occur prior to an energetic geologic event such as an earthquake, landslide, or volcanic eruption. Clear recognition of unambiguous premonitory phenomena would allow anticipatory action to be taken in order to reduce the loss of life and reduction of damage during a geologic catastrophe. (See figures 5.14 and 5.15.)

Premonitory (Precursor) Events: natural geologic features that, due to abrupt or unusual changes, signal the imminent onset of a sudden, devastating event such as an earthquake, volcanic eruption, or landslide. (See figures 5.14 and 5.15.)

Preplate Stage: the Hadean time interval in Earth history (post-accretion but pre–Archean time) typified by a terrestrial magma ocean, followed by turbulent or poorly organized mantle convective overturn. No crust survived this stage. (See figure 7.11A.)

Pressure Solution: the dissolution of grain boundary material in granular rocks due to the focusing of overburden (lithostatic) pressure along grain-grain point contacts during loading.

Primary Rocks: original additions to the crust of igneous material derived from the mantle.

Primitive Mantle: the original composition of the mantle (stony meteorite) subsequent to infall of the core, but prior to partial melting and removal of liquids to form the Earth's crust.

Proterozoic Era: the time interval between about 2.5 billion and 570 million years ago. (See figure 7.2.)

Protolith: the original rock prior to the change to the present configuration.

Proton: an atomic nuclear particle possessing approximately unit mass and a single positive charge. The number of protons, Z, in an atom define the atomic number characteristic of the element, as well as its chemical behavior.

Proven Reserve: demonstrated, quantitative evaluation of the amount of a reserve for a particular natural resource (see reserve). Usually proven reserves are documented through drilling, exploratory excavation, or actual mining.

Pull-Apart: releasing bend along a continental strike-slip and/or transform fault boundary, characterized by a mass deficiency. Generally the site of a topographic basin. (See figure 1.19.)

Pumice: glassy, low-density lava dominated by bubbles (expanding and escaped gas).

Pyroclastic Rock: rock formed dominantely by the accumulation of solid and liquid particles (called ejecta) thrown out of a volcano during eruption. (See figure 4.1A, B.) They are variably layered clastic sedimentary rocks consisting dominantly—or exclusively—of igneous fragments.

Pyroxene Granulite: a high-grade metamorphic rock characterized by intermediate calcium-sodium or calcic plagioclase, one or two pyroxenes, and cordierite at low pressures, or garnet at high pressures. (See figure 4.19.) Granulites are characteristic of the deep continental crust.

Pyroxenite: ultramafic rock consisting of either orthopyroxene (hypersthene), or clinopyroxene (augite), or both.

Quantum: a discrete amount of energy, used in reference to describe the electron-orbital levels surrounding the central nucleus of an atom.

Quartzofeldspathic Rock: rock containing abundant quartz and feldspar. Granite gneiss is a typical example.

Radiative Heat Transfer: transmission of thermal energy by electromagnetic wave propagation. It is most important at high temperatures where materials become incandescent.

Radioactive: an element that spontaneously, and at a fixed rate proportional to its abundance, is converted to other, lower atomic number elements plus atomic particles, liberating large amounts of energy in the process.

Radiometric Age: the quantitative age of a mineral or rock as determined by analytical measurement of an originally incorporated radioactive element, and the newly produced (postcrystallization) radiogenic daughter product(s). (See figure 7.1.)

Radius Ratio: the fraction defined by the quotient of the cation radius divided by the anion radius. In ionic crystal structures, the distance between neighboring cation and anion centers is the sum of the respective radii. The radius ratio determines the number of anions that will surround a central cation in a crystal structure. (See figure 3.9 and 3.10.)

Recumbent Fold (Nappe): a tightly appressed (squeezed) fold that has been overturned. Both limbs are inclined (dip downward) in the same direction.

Reflected Wave: vibrational energy bounced back from a planar discontinuity at a path angle equal to the angle of transmission incidence.

Refracted Wave: vibrational energy is propagated through a planar discontinuity into a new medium at an angle that in general is different from the angle of transmission incidence (angle i). The bending, or refraction angle (angle r) is related to the transmission velocities, V_i and V_r, and the angle of incidence by the equation: $V_r \sin i = V_i \sin r$.

Regional Metamorphic Rocks: metamorphic rocks that develop from preexisting rocks through the process of recrystallization on a regional scale, unrelated to local igneous phenomena such as pluton emplacement. (See figures 4.15, 4.16, and 4.17.) Deformation typically accompanies regional metamorphism.

Remanent (Remnant) Magnetism: alignment of magnetic domains in the constituent mineraloids and/or minerals of a rock coincident with the Earth's magnetic field at the time of formation of the rock.

Renewable Resource: a resource that is being continuously generated, such as biomass, tidal, hydroelectric, wind, geothermal, and solar power.

Repeat Periodicity: the distance in any chosen direction within a crystal structure required to produce an identical atomic arrangement.

Reserve: that portion of a natural resource that can be profitably extracted, using present-day methods.

Reservoir Rock: a porous rock whose interstices are occupied by oil and/or gas, or by some other fluid of interest. (See figure 6.8.)

Retorting: the process of heating, thermally maturing or distilling, and extracting substances, such as hydrocarbons originally residing in a source rock, such as oil shale.

Reverse Fault: a high-vertical-angle fault in which the upper block—or hanging wall—has moved up relative to the lower, footwall block. Horizontal shortening characterizes this type of movement.

Reversed Magnetic Polarity: the Earth's magnetic field opposite to that of the present, whereby the north-seeking end of a compass needle would point south. (See figure 2.11.)

Rhyolite: a highly silicic, alkali-rich lava typical of and confined to the continental crust. (See chapter 4 and Table 4.1 for description and classification.)

Richter Scale: a commonly used scale of quantitative measurement of vibrational energy released in an earthquake. The magnitude of the event is proportional to the linear deflection of the recording pen of a standard seismometer normalized for a specific distance from the hypocenter. The total energy of shaking increases about thirty-two times between successive numbers, hence a 7.7 magnitude quake is approximately ten times as energetic as a 7.0 Richter scale event.

Rifted Continental (Passive) Margin: also called an Atlantic-type margin. Formed due to the breakup of continental crust by the sea-floor spreading process. Although initiated as a divergent lithospheric plate junction, continued spreading removes the passive margin from the vicinity of the ridge axis and divergent plate boundary. Sediments are laid down on the thermally subsiding continental shelf, slope, and rise as a prism of undeformed strata. (See figure 4.12A.)

Rifting, Drifting, and Suturing Stage: the Phanerozoic time in Earth history, typified by very slow, laminar flow of enormous mantle convection cells and by the modern cycle of plate tectonics. (See figure 7.11.)

Rock: a naturally occurring coherent aggregate of grains of one or more minerals and/or mineraloids, having a common origin. To be dignified by the term rock, such aggregations should be present in units large enough to represent an important (mapable) part of the Earth.

Rock Cycle: the continuous or episodic reworking of pre-existing crustal and uppermost mantle rocks during the geochemical evolution of the lithosphere. (See figure 4.22.)

Roof Pendant: a large expanse of country rock surrounded by, and hanging down into, a plutonic igneous body. (See figure 4.4.)

Sandstone: a medium-grained (grain size between one sixteenth and two millimeters) clastic sedimentary rock formed by the accumulation of granular fragments of rocks and minerals, generally in a marine basin or lake, but much less commonly as subaerial dune deposits. (See chapter 4 for description and classification.)

Scarce Resource: a natural resource whose concentration in the reachable parts of the Earth is quite low; accordingly, deposits of economic value may be very low in the substance in question, but nevertheless must be significantly above the average terrestrial concentration. Gold is a good example of a scarce resource.

Schist: a fairly fine-grained, foliated metamorphic rock characterized by homogeneous compositions throughout. The foliation, caused by parallelism of constituent platy and prismatic mineral grains, is called schistosity. (See figure 4.13A, B.)

Seamount: a submerged mountain within an ocean basin. (See figure 2.8.)

Sea-Floor Spreading: that process whereby new oceanic crust–capped lithosphere is generated in the vicinity of an oceanic ridge. The latter marks the site of a rising column or sheet of hot, buoyant (convecting) mantle material. (See figures 2.6, 4.9, and 4.20.)

Secondary Rocks: reworked crustal lithologic units, including some igneous rocks, all sedimentary, and all metamorphic rocks. Secondary units all develop from a lithic precursor (protolith).

Sedimentary Breccia: a rock that consists chiefly of angular conglomerate pebbles.

Sedimentary Differentiation: the chemical fractionation of protoliths by sedimentary processes (such as weathering, abrasion, and solution during transportation, etc.) into product rocks greatly enriched in certain constituents; e.g., SiO_2, $CaCO_3$, MnO, Fe_2O_3, or Al_2O_3. (See figure 4.11.)

Sedimentary Rock: any rock formed at or near the surface of the Earth through the settling of material. Two broad classes are recognized: (1) materials chemically and organically precipitated from aqueous solutions (principally seawater) and (2) discrete grains (or clasts) or particulate matter (detritus) brought to a site of deposition by running water, moving ice, wind, or downhill slumpage.

Seismic Waves: vibrational energy mechanically transmitted through the Earth (a variety of sound waves). (See figure 1.8A, B.)

Seismograph: an instrument sited at a fixed location that records and amplifies the vibrations in specific directions of the Earth.

Shale: clastic sedimentary rock made up of constituent minerals possessing a grain size less than $\frac{1}{16}$ millimeters—generally less than $\frac{1}{256}$ millimeters.

Shallow-focus Earthquake: energy release (hypocenter) at depth within the Earth of zero to fifty kilometers.

Shear: differential motion between two masses of matter. For solids undergoing shear, deformation—or distortion of the material—may be elastic (recoverable) or plastic (permanent). If not, failure by rupture (faulting) may ensue.

Shear Waves, Seismic: so-called S waves, transmitted by an undulatory vibration transverse to the propagation direction, rather like the movement of energy along an oscillating rope or string with fixed ends. The waves are called "S" for "secondary," because they travel at lesser speeds through the Earth than primary waves, hence are received later by a seismograph station. (See figure 1.8B.)

Sheet Silicate: a group of minerals characterized by a layer of silicon-oxygen tetrahedral six-member rings polymerized to infinity in two directions. The unit repeat of the sheet is $(Si_4O_{10})^{-4}$, and each tetrahedron has three bridging oxygens and one nonbridging oxygen. (Of course, only half of each bridging oxygen can be assigned to a specific tetrahedron.) Some of the silicon is replaced by aluminum in certain minerals, such as the micas. (See figure 3.20.)

Shield Volcano: an igneous constructional feature, possessing a broad, low outline, chiefly built through the extrusion of rather fluid lava. (See figure 4.2.)

Sial: continental crust of the Earth, relatively rich in SiO_2, Al_2O_3, K_2O, Na_2O, volatiles, and radioactive elements. (See figure 1.7B.)

Silicic Igneous Rock: magmatic rock rich in silica, and generally also sodium and potassium, but poor in ferromagnesian constituents, iron, magnesium, and calcium.

Sill: a tabular igneous intrusive concordant (parallel) to the layering of the wall rocks. (See figure 4.5B.)

Siltstone: a slightly finer-grained equivalent of sandstone (grain size between $1/16$ and $1/256$ millimeters).

Sima: oceanic crust of the Earth, relatively rich in CaO, FeO, and MgO and poor in alkalies, volatiles and SiO_2. (See figure 1.7B.)

Single-Chain Silicate: a group of minerals characterized by a chain of silicon-oxygen tetrahedra polymerized to infinity along a single direction. The unit repeat of the chain is $(SiO_3)^{-2}$, and each tetrahedron possesses two bridging and two nonbridging oxygens. (Of course, only half of each bridging oxygen can be assigned to a specific tetrahedron.) (See figure 3.9A.)

Slate: a fine-grained metamorphic rock, the percursor of which was generally a mudstone. Numerous parallel fractures impart a cleavage to such rocks. (See chapter 4 for description and classification.)

Smelting (Roasting): heating an ore to high temperatures in order to drive off volatiles such as sulfur.

Soil Profile: zonation from surficial soil (dirt) downward to fresh rock: zone A is organic rich, leached of soluble constituents and unconsolidated; zone B is Al_2O- and Fe_2O_3-rich; zone C denotes fresh but fractured bedrock.

Solar Nebula: primordial cloud of gas, condensation of which resulted in formation of the Sun and the encircling planetary bodies, meteorites, asteroids, and comets of the peripheral Oort cloud. In aggregate, these condensates constitute our solar system.

Solar Wind (Solar Flux): radiation of electromagnetic energy from the Sun, including electrically charged gas (plasma) as well as light.

Solid Solution: chemical variation of a mineral series among several (two or more) end-member compositions. A good example is the plagioclase solid-solution series, specific mineral examples of which range from albite, $NaAlSi_3O_8$, to anorthite, $CaAl_2Si_2O_8$.

Specific Gravity: mass of an object divided by the mass of an equivalent volume of water (at four degrees Celcius).

Spreading Axis (Spreading Center, Oceanic Ridge, Divergent Plate Boundary): the locus where oceanic crust–capped lithospheric plates move orthogonally apart. Here is where mid-oceanic ridge basalts are extruded on the sea floor. (See figures 2.6, 4.9, and 4.20.)

Spreading Center: a linear zone along which new oceanic crust–capped lithosphere forms—generally called a divergent plate boundary. This zone is marked by an oceanic ridge. (See figures 2.6, 4.9, and 4.20.)

Stony Meteorite: any meteorite composed chiefly of silicate materials, normally magnesium-rich minerals such as olivine and/or enstatite.

Strain: the response of a substance—in the geologic case, a rock—to differential forces or stress. Strain includes brittle failure (fracture), elastic distortion, and ductile deformation.

Strata: sedimentary layers of rock, formed parallel to the original floor of the basin of deposition.

Strategic Mineral: a resource that is essential for the efficient running of a nation but that must be imported from abroad. (See figure 6.2.)

Stratigraphic Trap: lateral changes in a petroleum-bearing reservoir rock that cause the reservoir volume to decrease to zero thickness; this pinch-out causes a confinement of the oil- and gas-bearing horizon. (See figure 6.8D.)

Stratigraphy: the study of layered (sedimentary and volcanoclastic) rocks.

Stratovolcano: symmetric central cone, chiefly constructed through the venting of viscous lava and voluminous ash. (See figure 4.3.)

Stress: differential forces, or pressures, applied to a substance—in the geologic case, a rock.

Strike-Slip Fault: a continental fault characterized by differential horizontal motion parallel to the length of the fracture. Most large strike-slip faults are probably intracontinental transforms (e.g., see figures 5.3, 5.5, and 5.7).

Stromatolites: primitive, carbonate-secreting algae that build colonial mats and lagoonal incrustations in marine water. Their most abundant flourishing was in the Precambrian.

Structural Trap: folds and faults in which impermeable cap rocks overlie, or are juxtaposed against, a petroleum-bearing reservoir rock, preventing escape of the oil and gas. (See figure 6.8A, C.)

Subduction: that process whereby lithosphere is destroyed by descending into the deeper mantle in the vicinity of an oceanic trench. Generally, the sinking slab is relatively old, dense lithosphere capped by oceanic crust. (See figures 2.7, 4.10, and 4.21.)

Subjacent Pluton (Batholith, Stock, Cupola): downward-enlarging igneous body of irregular shape, at least partly discordant. Large ones (about 300 square kilometers) are called batholiths, successively smaller exposed bodies are termed stocks and cupolas. (See figure 4.4.)

Subsilicic Igneous Rock: magmatic rock poor in silica, generally rich in the ferromagnesian constituents iron, magnesium, and calcium.

Substitution Solid Solution: a solid-solution series in which chemical variation is described as replacement of one (or more) ion by another of similar properties. This is the most common type of solid solution in minerals. As an example, the hypersthenes are intermediate Fe^{+2} plus magnesium orthopyroxenes whose compositions may be derived from the magnesian end member enstatite, $MgSiO_3$, by the substitution of ferrous iron. The chosen example involve cation exchange, but in some cases anion replacement (e.g., F^{-1} for OH^{-1}) is also possible.

Supercratonal Stage: the Proterozoic Era in Earth history, typified by laminar flow of middle-sized mantle convection cells. Supercontinental assemblies represent the accretion of Archean microcontinental and island-arc fragments, plus continental-crust generation. (See figure 7.11C.)

Superposition, Law of: for stratified rocks deposited at the Earth's surface, older units are near the bottom of a layered section, with the oldest at the base; progressively younger units occur farther upsection (upward).

Surface Waves: vibrational energy propagated along the surface of a body such as the Earth. Waves in oceans and lakes are also surface waves.

Symmetry: symmetry is that property of an object (such as a crystal) whereby a translation, rotation about an axis, or reflection across a mirror plane, results in self-coincidence (an identical aspect).

Syncline: a bowl or hammock-shaped fold in layered rocks in which the beds are inclined (dip downwards) towards the central low. (See figures 6.8B and 7.9.)

Synclinorium: a great, linear downwarp consisting of subparallel, subsidiary anticlines and dominant synclines. Basically a large syncline with superimposed minor anticlines and synclines (folding at two different scales).

Talus: the clastic debris that accumulates at the base of a steep slope, generally building up a cone of material toward the source area.

Tar Sand: a reservoir rock in which liquid hydrocarbons have been degraded by near-surface oxidation and devolatilization to viscous, semisolid asphalt or tar.

Terrane: a fault-bounded block of the Earth's crust, the lithologic units of which share common or related histories, but that in aggregate constrasts strongly with adjacent blocks. Because of their lack of genetic relationships, most terranes appear to be far-traveled, or exotic, compared with neighboring, juxtaposed terranes.

Tetrahedral Chain: the infinite polymerization of silicon-oxygen tetrahedra in one dimension. These linkages give rise to $(SiO_3)^{-2}$ (single-chain, characterized by two nonbridging oxygens) and $(Si_4O_{11})^{-6}$ (double-chain, characterized on the average by 1.5 nonbridging oxygens) silicate mineral groups. (See figure 3.19.) Small amounts of the silicon may be replaced by tetrahedrally coordinated aluminum.

Tetrahedral Framework: the infinite polymerization of silicon-oxygen tetrahedra in three dimensions. Gives rise to SiO_2-type silicates, in which all oxygens are bridging oxygens. (See figure 3.21.) Up to one half of the silicon may be replaced by tetrahedrally coordinated aluminum.

Tetrahedral Rings: three-, four- or six-member rings, having formulas for the structural subunits of $(Si_3O_9)^{-6}$, $(Si_4O_{12})^{-8}$, and $(Si_6O_{18})^{-12}$, respectively. In such configurations, each tetrahedron contains two bridging oxygens (each shared with another tetrahedron) and two nonbridging oxygens. (See figure 3.18C.)

Tetrahedral Sheet: the infinite polymerization of silicon-oxygen tetrahedra in two dimensions. Gives rise to $(Si_4O_{10})^{-4}$ sheet-silicates. Each tetrahedron has but one nonbridging oxygen. (See figure 3.20.) About one fourth of the silicon may be replaced by tetrahedrally coordinated aluminum.

Therapsid Reptiles: distinctive, latest Paleozoic mammal-like reptiles geographically confined to the Gondwana supercontinent. (See figure 2.3—cute little rascals, aren't they?)

Thermal Gradient: any change of temperature with distance. The increase of temperature with depth within the Earth is called the geothermal gradient, and is usually given in units of degrees Celcius per kilometer. (See figures 1.10 and 7.3.)

Thermal Maturation: the distillation process that converts low-grade organic matter to liquid and gaseous hydrocarbons during heating by driving off volatile constituents.

Thrust Fault: a low-vertical-angle (about fifteen to twenty degrees) fault in which the upper block—or hanging wall—has moved relatively up and over the lower block; it often places older rocks over younger and is characteristic of accretionary prisms and subduction zones. (See figures 2.7, 4.10, and 4.21.) Thrust faulting results in lateral compression of the section.

Transform Fault: a special type of fault, the active portion of which links segments of divergent and/or convergent lithospheric plate boundaries. (See figures 2.9, 2.14, 5.3, 5.5, and 5.7.)

Transition Zone: that part of the Earth's mantle sited between the upper and lower mantle, at depths of about 400–670 kilometers. Here physical properties change rapidly but continuously with depth. (See figure 1.7A, B.)

Trench Basin: the linear bathymetric low associated with a downturning oceanic crust–capped lithospheric plate. The trench marks the surface expression of the subduction zone, or inclined convergent plate junction. (See figure 4.12B, C.)

Trench Complex: the rocks and rocky debris that accumulates adjacent to or within the confines of an oceanic trench. Because of convergent and oblique plate motion, material originating some distance away from the trench may ultimately be brought into this site, decoupled from the downgoing—or subducting—plate, and accreted to the stable, nonsubducted slab by shallow offscraping or deeper level underplating. (See figures 2.7, 4.10, and 4.21.)

Triple Junction: the surface point common to three lithospheric plates. Several types are common—among them, ridge-ridge-ridge, and ridge-trench transform. Such junctions may evolve rapidly with time due to differential plate motions. (See figures 5.4A, B, C and 5.5.)

Tsunami: popularly termed a tidal wave, but more accurately, a seismic sea wave. Such waves are caused by earth movements accompanying earth quakes, less commonly, volcanic eruptions and/or landslides.

Tuff: a pyroclastic deposit consisting chiefly of volcanic ash (see ejecta).

Twinned Crystals: composite crystals consisting of two (or more) portions that have identical geometries but contrasting orientations. Simple twins consist of two parts, whereas multiple twins consist of many individual crystal segments. Twinning accounts for the striations observed on plagioclase. (See figure 3.12H.)

Ultramafic Rock: a lithologic unit richer in ferromagnesian constituents, iron, magnesium, often calcium, and poorer in alkalies and silica than mafic rocks. Ultramafic is often used nearly synonymously with "peridotite" and "mantle material."

Ultrametamorphism: the partial melting of metamorphic rock to produce layered (mafic) refractory solid and more fusible (quartzofeldspathic) molten lithologic materials—in aggregate termed migmatites. (See figure 4.10.)

Uniformitarianism, Hypothesis of: the theory that the types of geologic processes operating today attended the Earth in the distant geologic past.

Unit Cell: the atomic repeat periodicity along the three principal crystallographic axes of a crystalline material. Mineral grains and crystals are edifices constructed by the combination of unit cells in exactly the same orientation, leaving neither gaps nor allowing overlaps. (See figures 3.15 and 3.16.)

Upper Mantle: the outer portion of the Earth's mantle extending from the base of the crust to depths of 400 ± 50 kilometers. (See figure 1.7A, B.)

Valence Electron(s): outer-orbital electron(s) that may be removed to form a positively charged ion (cation), or added to form a negatively charged ion (anion); both types of stable ion (+ and −) possess the noble gas electronic configuration (filled orbitals).

Van Der Waals Bonding: very, very weak or "residual" attractive forces resulting from the nonspherical distribution of positive and negative charges between bonded atoms. (See figure 3.13B.)

Vein Deposit: the result when hydrothermal solutions that are cooling or encountering changing mineralogic and/or chemical conditions pass through fractures in rocks, characteristically reacting with their surroundings and precipitating out minerals. Whether the mineralogic aggregate is economic or not, the deposit within the fracture is termed a vein.

Viscous Drag: the entrainment of buoyant material on a sinking, relatively dense slab because of structural coherence and integrity of the entire unit. The concept generally evokes downward movement of the crust-capped lithosphere above, resulting from coupling to flowing asthenosphere below. (See figure 7.11.)

Volatile Phase: a gas, consisting of one or more highly volatile constituents (e.g., H_2O, CO_2, H_2S).

Volcanic Breccia: a rock consisting chiefly of angular fragments of lava, pumice, conduit walls, etc. encased in lava or a matrix of tuff.

Volcanic Neck: a pipelike or fissurelike conduit filled with igneous rock, thought to represent the throat of a volcano. (See figure 4.6.)

Volcanic/Plutonic Arc: an island arc or continental margin characterized by voluminous volcanic rocks and their deep-seated, intrusive equivalents. Andesite is the characteristic—but not exclusive—rock type. (See figures 2.7, 4.10, and 4.21.)

Wave Base: that ocean depth (usually a few meters) below which wave action, and thus wave-related erosion, do not exist. Deep-sea currents do occur, both along continental slopes and down submarine canyons.

Weathering: the mechanical and chemical attack of pre-existing materials under near-surface geologic conditions.

Zeolite Facies Rock: a very low-grade metamorphic rock characterized by zeolite minerals such as laumontite, analcime, quartz, clay minerals, and hydrous ferromagnesian phases. (See figure 4.19.) Such metamorphic assemblages are produced in the near-surface environment.

=REFERENCES FOR ILLUSTRATIONS
=AND TABLES

Allaart, J. H. 1976. The pre–3760 Myr-old supracrustal rocks of the Isua area, central West Greenland and the associated occurrence of quartz banded ironstone. *In* B. F. Windley, ed., *The Early History of the Earth*, pp. 177–189. London: Wiley.

Anderson, D. L. 1971. The San Andreas fault. *Scientific American*, 225 (5):52–68.

Anhaeusser, C. R. 1971. The Barberton Mountain Land, South Africa: A Guide to the Understanding of the Archaean Geology of western Australia. In J. E. Glover, ed., *Symposium on Archean Rocks*, Geol. Soc. Australia Spec. pub. no. 3, pp. 103–119.

Anhaeusser, C. R., R. Mason, M. J. Viljoen, and R. P. Viljoen. 1969. A reappraisal of some aspects of Precambrian shield geology. *Geol. Soc. Amer. Bull.*, 80:2175–2200.

Averitt, P. 1975. Coal resources of the United States. *U.S. Geological Survey Bulletin 1412.*

Barazangi, M. and J. Dorman. 1969. World seismicity map compiled from ESSA Coast and Geodetic Survey epicenter data, 1961–1967. *Seismological Soc. America Bull.*, 59:369–80.

Bullard, E. C., J. E. Everett and A. G. Smith. 1965. The fit of the continents around the Atlantic. *Phil. Trans. Roy. Soc. London*, A258:41–51.

Clarke, F. W. 1924. Data of geochemistry. *U.S. Geological Survey Bulletin 770.*

Condie, K. C.. 1982. *Plate Tectonics and Crustal Evolution*. New York: Pergamon Press.

Decade of North American Geology. 1983. *Geologic Time Scale*. Boulder, Colo.: Geol. Soc. America.

Dennen, W. H. 1960. *Principles of Mineralogy*. New York: N.Y.: Ronald Press.

Dewey, J. F. 1972. Plate Tectonics. *Scientific American*, 226 (5):56–68.

Dickinson, W. R., R. V. Ingersoll, D. S. Cowan, K. P. Helmold, and C. A. Suczek. 1982. Provenance of Franciscan graywackes in coastal California. *Geol. Soc. America Bull.*, 93:95–107.

Dietz, R. S. and J. C. Holden. 1970. Reconstruction of Pangaea: Breakup and dispersion of continents, Permian to present. *Jour. Geophys. Research*, 75:4939–56.

Engebretson, D. C. A. Cox, and R. G. Gordon. 1985. Relative motion between oceanic and continental plates in the Pacific Basin. *Geol. Soc. America Special Paper 206.*

Ernst, W. G. 1969. *Earth Materials.* Englewood Cliffs, N.J.: Prentice-Hall.

Ernst, W. G. 1976. *Petrologic Phase Equilibria.* San Francisco: Freeman.

Ernst, W. G. 1979. California and plate tectonics, an interpretive account. *California Geology*, 32:187–96.

Ernst, W. G. 1983. A summary of Precambrian crustal evolution. In S. J. Boardman, ed., *Revolution in the Earth Sciences*, pp. 36–55. Dubuque, Iowa: Kendal-Hunt.

Francheteau, J. 1983. The oceanic crust. *Scientific American*, 249 (13):114–29.

Hartmann, W. K. 1983. *Moons and Planets*, 2d ed. Belmont, Calif.: Wadsworth.

Head, J. W. and S. C. Solomon 1981. Tectonic evolution of the terrestrial planets. *Science*, 213 (4503):62–76.

Heezen, B. C. and M. Tharp. 1977. World ocean floor. Map, U.S. Navy, scale 1:23,230,300.

Heirtzler, J. R. et al. G. O. Dickinson, E. M. Herran, W. C. Pitman III, and X. LePichon, 1968. Marine magnetic anomalies, geomagnetic field reversals, and motions of the ocean floor and continents. *Jour. Geophys. Res.*, 73:2119–36.

Heirtzler, J. R., X. LePichon, and J. G. Baron. 1966. Magnetic anomalies over the Reykjanes ridge. *Deep-Sea Research*, 13:427–43.

Hubbert, M. King. 1974. *U.S. Energy Resources: A Review as of 1972.* U.S. Senate, Committee on Interior and Insular Affairs, 93rd Cong., 2d sess., Serial No. 1 (93–140).

Hubbert, M. King. 1978. World resources of fossil organic raw materials. In L. E. St. -Pierre, ed., *Resources of Organic Matter for the Future, Perspectives and Recommendations: CHEMRAWN Conference I*, pp. 58–98. Toronto: Multiscience Publications.

Hubbert, M. King. 1985. The world's evolving energy system. In R. L. Perrine and W. G. Ernst, eds., *Energy: For Ourselves and Our Posterity*, pp. 44–100. Englewood Cliffs, N.J.: Prentice-Hall.

Hurley, Patrick M. 1968. The confirmation of continental drift. *Scientific American*, 218 (4):52–64.

Ingersoll, R. V. 1982. Triple-junction instability as cause for late Cenozoic extension and fragmentation of the western United States. *Geology*, 10:621–24.

Irving, E. and J. McGlynn. 1981. On the coherence, rotation and paleolatitude of Laurentia in the Proterozoic. In A. Kröner, ed., *Precambrian Plate Tectonics*, pp. 561–98. Amsterdam: Elsevier.

James, H. L. 1955. Zones of regional metamorphism in the Precambrian of Northern Michigan. *Geol. Soc. Amer. Bull.*, 66:1455–88.

Kieffer, S. W. 1989. Geologic nozzles. *Rev. Geophysics*, 27:3–38.

Leet, L. D., M. E. Kauffman, and S. Judson 1987. *Physical Geology*, 7th ed. Englewood Cliffs, N.J.: Prentice-Hall.

Mason, B. 1966. *Principles of Geochemistry.* New York: Wiley.

Pearce, J. A. 1976. Statistical analysis of major element patterns in basalts. *Jour. Petrology*, 17:15–43.

Pitman, W. D., R. L. Larson and E. M. Herron. 1974. *The Age of the Ocean Basins.* Geol. Soc. America Map Series. Boulder, Colo.

Poldervaart, A., 1955, Chemistry of the earth's crust. In A. Poldervaart, ed., *Geol. Soc. Amer. Special Paper 62*, pp. 119–44.

Press, F. and R. Siever. 1986. *Earth*, 4th ed. San Francisco: Freeman.

Ringwood, A. E. 1975. *Composition and Petrology of the Earth's Mantle*. New York: McGraw-Hill.

Seager, W. R. and P. Morgan. 1979. Rio Grande Rift in southern New Mexico, West Texas, and northern Chihuachua. In A. E. Riecker, ed., *Rio Grande Rift: Tectonics and Magmatism*, pp. 87–106 Washington, D.C.: American Geophysical Union.

Shannon, R. D. 1976. Revised effective ionic radii and systematic studies of interatomic distances in halides and chaleogenides. *Acta Cryst.*, A32:751–57.

Shannon, R. D. and C. T. Prewitt. 1969. Effective ionic radii in oxides and fluorides. *Acta Cryst.*, B25:925–46.

Silver, E. A. and R. B. Smith. 1983. Comparison of terrane accretion in modern Southwest Asia and the Mesozoic North American Cordillera. *Geology*, 11:198–202.

Simkin, T. and L. Siebert. 1984. Explosive eruptions in space and time: Deviations, intervals, and a comparison of the World's active volcanic belts. Studies in Geophysics. *Explosive Volcanism: Inception, Evolution, and Hazards*, pp. 110–21. Washington, D.C.: National Academy Press.

Skinner, B. J. 1986. *Earth Resources*. 3d ed. Englewood Cliffs, N.J.: Prentice-Hall.

Thorarinsson, Sigurdur 1964. Surtsey—*A New Island in the Atlantic. Almenna Bokafelagid.*, Reykjavik, Iceland:63p.

Toppozada, T. R., C. R. Real, and D. L. Parke. 1986. Earthquake history of California. *California Geology*, 39 (2):27–33.

Viljoen, R. P. and M. J. Viljoen. 1971. The geological and geochemical evolution of the Onverwacht volcanic group of the Barberton Mountain Land, South Africa. In J. E. Glover, ed., *Symposium on Archean Rocks*. Geol. Soc. Australia Spec. Pub. no. 3.

☰ INDEX

THE PERIODIC TABLE

Group / Period	I	II											III	IV	V	VI	VII	VIII
1	1 H Hydrogen																	2 He Helium
2	3 Li Lithium	4 Be Beryllium											5 B Boron	6 C Carbon	7 N Nitrogen	8 O Oxygen	9 F Fluorine	10 Ne Neon
3	11 Na Sodium	12 Mg Magnesium											13 Al Aluminum	14 Si Silicon	15 P Phosphorus	16 S Sulfur	17 Cl Chlorine	18 Ar Argon
4	19 K Potassium	20 Ca Calcium	21 Sc Scandium	22 Ti Titanium	23 V Vanadium	24 Cr Chromium	25 Mn Manganese	26 Fe Iron	27 Co Cobalt	28 Ni Nickel	29 Cu Copper	30 Zn Zinc	31 Ga Gallium	32 Ge Germanium	33 As Arsenic	34 Se Selenium	35 Br Bromine	36 Kr Krypton
5	37 Rb Rubidium	38 Sr Strontium	39 Y Yttrium	40 Zr Zirconium	41 Nb Niobium	42 Mo Molybdenum	43 Tc Technetium	44 Ru Ruthenium	45 Rh Rhodium	46 Pd Palladium	47 Ag Silver	48 Cd Cadmium	49 In Indium	50 Sn Tin	51 Sb Antimony	52 Te Tellurium	53 I Iodine	54 Xe Xenon
6	55 Cs Cesium	56 Ba Barium	57 La Lanthanum	72 Hf Hafnium	73 Ta Tantalum	74 W Tungsten	75 Re Rhenium	76 Os Osmium	77 Ir Iridium	78 Pt Platinum	79 Au Gold	80 Hg Mercury	81 Tl Thallium	82 Pb Lead	83 Bi Bismuth	84 Po Polonium	85 At Astatine	86 Rn Radon
7	87 Fr Francium	88 Ra Radium	89 Ac Actinium															

— Transition Metals —

— Metalloids and Nonmetals —

Lanthanides (Rare Earth Metals)

58 Ce Cerium	59 Pr Praseodymium	60 Nd Neodymium	61 Pm Promethium	62 Sm Samarium	63 Eu Europium	64 Gd Gadolinium	65 Tb Terbium	66 Dy Dysprosium	67 Ho Holmium	68 Er Erbium	69 Tm Thulium	70 Yb Ytterbium	71 Lu Lutetium

Actinides

90 Th Thorium	91 Pa Protoactinium	92 U Uranium	93 Np Neptunium	94 Pu Plutonium	95 Am Americium	96 Cm Curium	97 Bk Berkelium	98 Cf Californium	99 Es Einsteinium	100 Fm Fermium	101 Md Mendelevium	102 No Nobelium	103 Lw Lawrencium